Darwin and the Mysterious Mr. X

ALSO BY LOREN EISELEY

The Immense Journey, 1957
Darwin's Century, 1958
The Firmament of Time, 1960
The Unexpected Universe, 1969
The Invisible Pyramid, 1970
The Night Country, 1971
Notes of an Alchemist, 1972
The Man Who Saw Through Time, 1973
The Innocent Assassins, 1973
All the Strange Hours, 1975
Another Kind of Autumn, 1977
The Star Thrower, 1978

Darwin and the

Illustrated with photographs

LOREN EISELEY

Mysterious Mr. X

NEW LIGHT ON
THE EVOLUTIONISTS

A Harvest/HBJ Book
Harcourt Brace Jovanovich
New York and London

Printed in the United States of America

Harvest/HBJ edition published by arrangement with E. P. Dutton

Library of Congress Cataloging in Publication Data
Eiseley, Loren C 1907-1977.
Darwin and the mysterious Mr. X.
(A Harvest/HBJ book)
Includes bibliographical references and index.
1. Darwin, Charles Robert, 1809-1882.
2. Blyth, Edward, 1810-1873. 3. Evolution—History.
4. Naturalists—England—Biography. I. Title.
[QH31.D2E38 1981] 575'.0092'4 80-24833
ISBN 0-15-623949-3

First Harvest/HBJ edition 1981

A B C D E F G H I J

Contents

Editor's Preface

A LITERARY CRITIC once made the observation that if Loren Eiseley had lived in the nineteenth century, he would have been a novelist.

The Immense Journey and other books by this naturalist, who died in 1977, attest to his gifts as a creative writer. But Eiseley, who lived in the Age of Science, chose to be an anthropologist—a student of man. In one sense, he combined his interests, for the novelist, who portrays characters and actions professing to be those of real life, must also be a student of man.

The University of Pennsylvania honored Eiseley with a special Chair, thus recognizing his unique position as a humanist–scientist. As Benjamin Franklin Professor of Anthropology and the History of Science, his academic duties were intentionally light in order to give him time to pursue his writing career.

Eiseley's literary accomplishments may have overshadowed his scientific endeavors and often confused those who could not understand what errand he was on. After all, why was he, a scientist, elected to the National Institute of Arts and Letters and a recipient of many of the major literary awards?

It is, however, possible for a man to be two things at the

same time, as his most remarkable discovery, told about in the pages of this book, reveals.

The present work is best described as a science book, for it is not anecdotal and personal like many of Eiseley's more literary writings. With one exception, these pieces were written for scientific and scholarly publications or prepared to be delivered before a gathering of scientists. Nevertheless, they are gracefully written and often poetic, for Eiseley was incapable of writing a dull or inelegant sentence.

"Charles Darwin," the first essay, recounts how in 1831 this complex Englishman set forth on his famous voyage in the *Beagle* and how after many years he published the *Origin of Species,* which revolutionized man's view of nature and his place in it. In "Alfred Russel Wallace," Eiseley tells how another great nineteenth-century naturalist and Darwin simultaneously announced the theory of evolution by means of natural selection. This is followed by a biography of Charles Lyell. Lyell founded modern geology and set the stage for the achievement of Darwin, but, until late in his career, he was reluctant to accept the idea of evolution.

The next essay, "Charles Darwin, Edward Blyth, and the Theory of Natural Selection," is what sets this book apart from every other written on the evolutionists. It is a study of the forgotten work of an English naturalist, Edward Blyth. Here Eiseley argues that the basic tenets of the theory of natural selection were stated by Blyth much earlier than by Darwin, and that Darwin was familiar with Blyth's idea, made use of it in his theories, and yet failed to acknowledge his obligation to Blyth.

This is Eiseley's discovery. He elaborates on his arguments and provides additional evidence in "Darwin, Coleridge, and the Theory of Unconscious Creation." Blyth's articles, which are followed here by a memoir of the naturalist, enable a reader to delve into the documentary evidence Eiseley uncovered; without them, his conclusions would be almost incredible.

The last part of the volume deals with human evolution, "the highest and most interesting problem for the naturalist," in the words of Darwin. "Neanderthal Man and the Dawn of

Human Paleontology" tells how the discovery of the first human fossil was received over a hundred years ago. The world wanted to hear what the author of *Origin of Species* had to say on the evolution of man, and an assessment of the significance of *The Descent of Man,* Darwin's work on the subject, follows. "The Time of Man" brings the story of our species forward into the twentieth century. Today new finds are revolutionizing our ideas about man's origins and the growth of that unique great brain which he may yet put to fatal use.

To return to the central essay of the book, although we have foreknowledge of the outcome of the story of Darwin and Blyth, this does not make the tale any less interesting. Eiseley has brought to light historical documents that absorb one's attention, and the detective-like analysis he uses to prove that Darwin was familiar with the articles of Blyth is fascinating in itself.

Indeed, the image of Eiseley as novelist emerges here again, for he first thought of describing his discovery in the form of a detective story, "The Case of Charles Darwin and the Mysterious Mr. X," from which the title of the present book is taken. Several pages of this manuscript exist, but Eiseley likely had second thoughts and realized that if he were to rescue from oblivion the memory of this remarkable man, Blyth, he would have to publish his account in a scholarly journal.

The essays in this volume were published over a period of time. The piece on Charles Darwin, for example, appeared several years before the Blyth essay and contains no mention of this "forgotten parent," obviously because Eiseley had not yet made his discovery. Because the essays were written at different times for different purposes, there is a certain amount of overlapping, and they have been edited accordingly.

The notes at the back of the book are of two types: in some cases they continue or amplify discussion in the text; in others they merely document statements. The acknowledgments are not only an expression of thanks for the use of material but also provide additional bibliographical information.

Eiseley's many fans, as well as students, historians of science and biologists, will find *Darwin and the Mysterious Mr. X* rewarding reading. But for Mabel Eiseley, the author's widow, and Caroline E. Werkley, his assistant, who made the papers available to the editor and consulted with him, this book would not have been possible.

KENNETH HEUER

No one can take from us the joy of the first becoming aware of something, the so-called discovery. But if we also demand the honor, it can be utterly spoiled for us, for we are usually not the first. *What does discovery mean, and who can say that he has discovered this or that? After all it's pure idiocy to brag about priority, for it's simply unconscious conceit, not to admit frankly that one is a plagiarist.*

—GOETHE

PART

The Dancers in the Ring

Sir Francis Bacon, the English philosopher and author, once spoke of those drawn into some powerful circle of thought as "dancing in little rings like persons bewitched." Our scientific models do simulate a sort of fairy ring, which, once it has encircled us, is hard to view objectively. In Charles Darwin's youth, the magic circle of fixity and that of organic novelty began to interpenetrate. The dancers bewitched by stable form discovered a new truth: evolution.

Charles Darwin

I

IN THE AUTUMN of 1831 the past and the future met and dined in London—in the guise of two young men who little realized where the years ahead would take them. One, Robert Fitzroy, was a sea captain who at twenty-six had already charted the remote, sea-beaten edges of the world and now proposed another long voyage. A religious man with a strong animosity toward the new-fangled geology, Captain Fitzroy wanted a naturalist who would share his experience of wild lands and refute those who used rocks to promote heretical whisperings. The young man who faced him across the table hesitated. Charles Darwin, four years Fitzroy's junior, was a gentleman idler after hounds who had failed at medicine and whose family, in desperation, hoped he might still succeed as a country parson. His mind shifted uncertainly from fox hunting in Shropshire to the thought of shooting llamas in South America. Did he really want to go? While he fumbled for a decision and the future hung irresolute, Captain Fitzroy took command.

"Fitzroy," wrote Darwin later to his sister Susan, "says the stormy sea is exaggerated; that if I do not choose to remain with them, I can at any time get home to England; and that if I like, I shall be left in some healthy, safe and nice country; that I shall always have assistance; that he has many books, all instruments, guns, at my service. . . . There is indeed a tide in the affairs of men, and I have experienced it. Dearest Susan, Goodbye."

They sailed from Devonport December 27, 1831, in H.M.S. *Beagle,* a ten-gun brig. Their plan was to survey the South American coastline and to carry a string of chronometrical measurements around the world. The voyage almost ended before it began, for they at once encountered a violent storm. "The sea ran very high," young Darwin recorded in his diary, "and the vessel pitched bows under and suffered most dreadfully; such a night I never passed, on every side nothing but misery; such a whistling of the wind and roar of the sea, the hoarse screams of the officers and shouts of the men, made a concert that I shall not soon forget." Captain Fitzroy and his officers held the ship on the sea by the grace of God and the cat-o'-nine-tails. With an almost irrational stubbornness Darwin decided, in spite of his uncomfortable discovery of his susceptibility to seasickness, that "I did right to accept the offer." When the *Beagle* was buffeted back into Plymouth Harbor, Darwin did not resign. His mind was made up. "If it is desirable to see the world," he wrote in his journal, "what a rare and excellent opportunity this is. Perhaps I may have the same opportunity of drilling my mind that I threw away at Cambridge."

So began the journey in which a great mind untouched by an old-fashioned classical education was to feed its hunger upon rocks and broken bits of bone at the world's end, and eventually was to shape from such diverse things as bird beaks and the fused wing-cases of island beetles a theory that would shake the foundations of scientific thought in all the countries of the earth.

II

The intellectual climate from which Darwin set forth on his historic voyage was predominantly conservative. Insular England had been horrified by the excesses of the French Revolution and was extremely wary of emerging new ideas which it attributed to "French atheists." Religious dogma still held its powerful influence over natural science. True, the seventeenth-century notion that the world had been created in 4004 B.C. was beginning to weaken in the face of naturalists' studies of the rocks and their succession of life forms. But the conception of a truly ancient and evolving planet was still unformed. No one could dream that the age of the earth was as vast as we now know it to be. And the notion of a continuity of events—of one animal changing by degrees into another—seemed to fly in the face not only of religious beliefs but also of common sense. Many of the greatest biologists of the time—men like Louis Agassiz and Richard Owen—tended to the belief that the successive forms of life in the geological record were all separate creations, some of which had simply been extinguished by historic accidents.

Yet Darwin did not compose the theory of evolution out of thin air. Like so many great scientific generalizations, the theory with which his name is associated had already had premonitory beginnings. All of the elements which were to enter into the theory were in men's minds and were being widely discussed during Darwin's college years. His own grandfather, Erasmus Darwin, who died seven years before Charles was born, had boldly proposed a theory of the "transmutation" of living forms. Jean Baptiste Lamarck had glimpsed a vision of evolutionary continuity. And Sir Charles Lyell—later to become Darwin's confidant—had opened the way for the evolutionary point of view by demonstrating that the planet must be very old—old enough to allow extremely slow organic change. Lyell dismissed the notion of catastrophic extinction of animal forms on a worldwide scale as impossible, and he made plain that natural forces—the work of wind and frost and water—were sufficient to explain most of the phenomena found in the rocks, provided these forces were seen as operating over enormous periods. Without Lyell's gift of time in immense quantities, Darwin would not have been able to devise the theory of natural selection.

If all the essential elements of the Darwinian scheme of nature were known prior to Darwin, why is he accorded so important a place in biological history? The answer is simple: Almost every great scientific generalization is a supreme act of creative synthesis. There comes a time when an accumulation of smaller discoveries and observations can be combined in some great and comprehensive view of nature. At this point the need is not so much for increased numbers of facts as for a mind of great insight capable of taking the assembled information and rendering it intelligible. Such a synthesis represents the scientific mind at its highest point of achievement. The stature of the discoverer is not diminished by the fact that he has slid into place the last piece of a tremendous puzzle on which many others have worked. To finish the task he must see correctly over a vast and diverse array of data.

Still it must be recognized that Darwin came at a fortunate time. The fact that another man, Alfred Russel Wallace, conceived the Darwinian theory independently before Darwin published it shows clearly that the principle which came to be called natural selection was in the air—was in a sense demanding to be born. Darwin himself pointed out in his autobiography that "innumerable well-observed facts were stored in the minds of naturalists ready to take their proper places as soon as any theory which would receive them was sufficiently explained."

III

Darwin, then, set out on his voyage with a mind both inquisitive to see and receptive to what he saw. No detail was too small to be fascinating and provocative. Sailing down the South American coast, he notes the octopus changing its color angrily in the waters of a cove. In the dry arroyos of the pampas he observes great bones and shrewdly seeks to relate them to animals of the present. The local inhabitants insist that the fossil bones grew after death, and also that certain rivers have the power of "changing small bones into large." Everywhere men wonder, but they are deceived through their thirst for easy explanations. Darwin, by contrast, is a working dreamer. He rides, climbs, spends long days on the Indian-haunted pampas in constant peril of his life. Asking at a house whether robbers are numerous, he receives the cryptic reply: "The thistles are not up yet." The huge thistles, high as a horse's back at their full growth, provide ecological cover for bandits. Darwin notes the fact and rides on. The thistles are overruning the pampas; the whole aspect of the vegetation is altering under the impact of man. Wild dogs howl in the brakes; the common cat, run wild, has grown large and fierce. All is struggle, mutability, change. Staring into the face of an evil relative of the rattlesnake, he observes a fact "which appears to me very curious and instructive, as showing how every character, even though it may be in some degree independent of structure . . . has a tendency to vary by slow degrees."

He pays great attention to strange animals existing in difficult environments. A queer little toad with a scarlet belly he whimsically nicknames *diabolicus* because it is "a fit toad to preach in the ear of Eve." He notes it lives among sand dunes under the burning sun, and unlike its brethren, cannot swim. From toads to grasshoppers, from pebbles to mountain ranges, nothing escapes his attention. The wearing away of stone, the downstream travel of rock fragments and boulders, the great crevices and upthrusts of the Andes, an earthquake—all confirm the dynamic character of the earth and its great age.

Captain Fitzroy by now is anxious to voyage on. The sails are set. With the towering Andes on their right flank they run north for the Galápagos Islands, lying directly on the Equator 600 miles off the west coast of South America. A one-time refuge of buccaneers, these islands are essentially chimneys of burned-out volcanoes. Darwin remarks that they remind him of huge iron foundries surrounded by piles of waste. "A little world in itself," he marvels, "with inhabitants such as are found nowhere else." Giant armored tortoises clank through the undergrowth like prehistoric monsters, feeding upon the cacti. Birds in this tiny Eden do not fear men: "One day a mocking bird alighted on the edge of a pitcher which I held in my hand. It began very quietly to sip the water, and allowed me to lift it with the vessel from the ground." Big sea lizards three feet long drowse on the beaches, and feed, fantastically, upon the seaweed. Surveying these "imps of darkness, black as the porous rocks over which they crawl," Darwin is led to comment that "there is no other quarter of the world, where this order replaces the herbivorous mammalia in so extraordinary a manner."

Yet only by degrees did Darwin awake to the fact that he had stumbled by chance into one of the most marvelous evolutionary laboratories on the planet. Here in the Galápagos was a wealth of variations from island to island—among the big tortoises, among plants and especially among the famous finches with remarkably diverse beaks. Dwellers on the islands, notably Vice Governor Lawson, called Darwin's attention to these strange variations, but as he confessed later, with typical Darwinian lack of pretense, "I did not for some time pay sufficient attention to this statement."

As one surveys the long and tangled course that led to Darwin's great discovery, one cannot but be struck by the part played in it by oceanic islands.

Until Darwin turned his attention to them, it appears to have been generally assumed that island plants and animals were simply marooned evidences of a past connection with the nearest continent. Darwin, however, noted that whole classes of continental life were absent from the islands; that certain plants which were herbaceous (nonwoody) on the mainland had developed into trees on the islands; that island animals often differed from their counterparts on the mainland.

Above all, the fantastically varied finches of the Galápagos amazed and puzzled him. There were parrot-beaks, curved beaks for probing flowers, straight beaks, small beaks—beaks for every conceivable purpose. These beak variations existed nowhere but on the islands; they must have evolved there. Darwin had early observed: "One might really fancy that, from an original paucity of birds in this archipelago, one species had been taken and modified for different ends." The birds had become transformed, through the struggle for existence on their little islets, into a series of types suited to particular environmental niches where, properly adapted, they could obtain food and survive. As the ornithologist David Lack has remarked, "Darwin's finches form a little world of their own, but one which intimately reflects the world as a whole."

Darwin's recognition of the significance of this miniature world, where the forces operating to create new beings could be plainly seen, was indispensable to his discovery of the origin of species. The island worlds reduced the confusion of continental life to more simple proportions; one could separate the factors involved with greater success. Over and over Darwin emphasized the importance of islands in his thinking. Nothing would aid natural history more, he contended to Lyell, "than careful collecting and investigating of *all the productions* of the most isolated islands. . . . Every sea shell and insect and plant is of value from such spots."

Darwin was born in precisely the right age even in terms of the great scientific voyages. A little earlier, the story the islands had to tell could not have been read; a little later much of it began to be erased. Today all over the globe the populations of these little worlds are vanishing, many without ever having been seriously investigated. Man, breaking into their isolation, has brought with him cats, rats, pigs, goats, weeds and insects from the continents. In the face of these hardier, tougher, more aggressive competitors, the island faunas—the rare, the antique, the strange, the beautiful—are vanishing without a trace. The giant Galápagos tortoises are almost extinct, as is the land lizard with which Darwin played. Some of the odd little finches and rare plants have gone or will go. On the island of Madagascar our own remote relatives, the lemurs, which have radiated into many curious forms, are now being exterminated through the destruction of the forests. Even that continental island Australia is suffering from the decimation wrought by man. The Robinson Crusoe worlds where small castaways could create existences idyllically remote from the ravening slaughter of man and his associates are about to pass away forever. Every such spot is now a potential air base where the cries of birds are drowned in the roar of jets, and the crevices once frequented by bird life are flattened into the long runways of the bombers. All this would not have surprised Darwin, one would guess.

IV

When Darwin reached home after the voyage of the *Beagle*, he was an ailing man, and he remained so to the end of his life. Today we know that this illness was in some degree psychosomatic, that he was anxiety-ridden, subject to mysterious headaches and nausea. Shortly after his voyage Darwin married his cousin Emma Wedgwood, granddaughter of the founder of the great pottery works, and isolated himself and his family in the little village of Down, in Kent. He avoided travel, save for brief trips to watering places for his health. For twenty-two years after the *Beagle*'s return he published not one word beyond the bare journal of his trip (later titled *A Naturalist's Voyage around the World*) and technical monographs on his observations.

Darwin's gardener is said to have responded once to a visitor who inquired about his master's health: "Poor man, he just stands and stares at a yellow flower for minutes at a time. He would be better off with something to do." Darwin's work was of an intangible nature which eluded people around him. Much of it consisted in just such standing and staring as his gardener reported. On a visit to the Isle of Wight he watched thistle seed wafted about on offshore winds and formulated theories of plant dispersal. Sometimes he engaged in activities which his good wife must surely have struggled to keep from reaching the neighbors. When a friend sent him a half ounce of locust dung from Africa, Darwin triumphantly grew seven plants from the specimen. "There is no error," he assured Lyell, "for I dissected the seeds out of the middle of the pellets." To discover how plant seeds traveled, Darwin would go all the way down a grasshopper's gullet, or worse, without embarrassment. His eldest son Francis spoke amusedly of his father's botanical experiments: "I think he personified each seed as a small demon trying to elude him by getting into the wrong heap, or jumping away all together; and this gave to the work the excitement of a game."

The point of his game Darwin kept largely to himself, waiting until it should be completely finished. He piled up vast stores of data and dreamed of presenting his evolution theory in a definitive, monumental book, so large that it would certainly have fallen dead and unreadable from the press. In the meantime, Robert Chambers, a bookseller and journalist, wrote and brought out anonymously a modified version of Lamarckian evolution, under the title *Vestiges of the Natural History of Crea-*

tion. Amateurish in some degree, the book drew savage onslaughts from the critics, including Thomas Huxley, but it caught the public fancy and was widely read. It passed through numerous editions in both England and America—evidence that *sub rosa* there was a good deal more interest on the part of the public in the "development hypothesis," as evolution was then called, than the fulminations of critics would have suggested.

Throughout this period Darwin remained stonily silent. Many explanations of his silence have been ventured by his biographers: that he was busy accumulating materials; that he did not wish to affront Fitzroy; that the attack on the *Vestiges* had intimidated him; that he thought it wise not to write upon so controversial a subject until he had first acquired a reputation as a professional naturalist of the first rank. A primary reason lay in his personality—a nature reluctant to face the storm that publication would bring about his ears. It was pleasanter to procrastinate, to talk of the secret to a few chosen companions such as Lyell and the great botanist Joseph Hooker.

The Darwin family had been well-to-do since the time of grandfather Erasmus. Charles was independent, in a position to devote all his energies to research and under no academic pressure to publish in haste.

"You will be anticipated," Lyell warned him. "You had better publish." That was in the spring of 1856. Darwin promised, but again delayed. We know that he left instructions for his wife to see to the publication of his notes in the event of his death. It was almost as if present fame or notoriety were more than he could bear. At all events he continued to delay, and this situation might very well have continued to the end of his life, had not Lyell's warning suddenly come true and broken his pleasant dream.

Alfred Russel Wallace, a comparatively unknown, youthful naturalist, had divined Darwin's great secret in a moment of fever-ridden insight while on a collecting trip in Indonesia. He, too, had put together the pieces and gained a clear conception of the scheme of evolution. Ironically enough, it was to Darwin, in all innocence, that he sent his manuscript for criticism in June of 1858.

Darwin, understandably shaken, turned to his friends Lyell and Hooker, who knew the many years he had been laboring upon his *magnum opus.* The two distinguished scientists arranged for the delivery of a short summary by Darwin to accompany Wallace's paper before the Linnean Society. Thus the theory was announced by the two men simultaneously.

The papers drew little comment at the meeting but set in motion a mild undercurrent of excitement. Darwin, though upset by the death of his son Charles, went to work to explain his views more fully in a book. Ironically he called it *An Abstract of an Essay on the Origin of Species* and insisted it would be only a kind of preview of a much larger work. Anxiety and devotion to his great hoard of data still possessed him. He did not like to put all his hopes in this volume, which must now be written at top speed. He bolstered himself by references to the "real" book—that Utopian volume in which all that could not be made clear in his abstract would be clarified.

His timidity and his fears were totally groundless. When the *Origin of Species* (the title distilled by his astute publisher from Darwin's cumbersome and half-hearted one) was published in the fall of 1859, the first edition was sold in a single day. The book which Darwin had so apologetically bowed into existence was, of course, soon to be recognized as one of the great books of all time. It would not be long before its author would sigh happily and think no more of that huge, ideal volume which he had imagined would be necessary to convince the public. The public and his brother scientists would find the *Origin* quite heavy going enough. His book to end all books would never be written. It did not need to be. The world of science in the end could only agree with the sharp-minded Huxley, whose immediate reaction upon reading the *Origin* was: "How extremely stupid not to have thought of that!" And so it frequently seems in science, once the great synthesizer has done his work. The ideas were not new, but the synthesis was. Men would never again look upon the world in the same manner as before.

No great philosophic conception ever entered the world more fortunately. Though it is customary to emphasize the religious and scientific storm the book aroused—epitomized by the famous debate at Oxford between Bishop Wilberforce and Thomas Huxley—the truth is that Darwinism found relatively easy acceptance among scientists and most of the public. The way had been prepared by the long labors of Lyell and the wide popularity of Chambers' book, the *Vestiges.* Moreover, Darwin had won the support of Hooker and of Huxley, the most formidable scientific debater of all time. Lyell, though more cautious, helped to publicize Darwin and at no time attacked him. Asa Gray, one of America's leading botanists, came to his defense. His co-discoverer, Wallace, generously advanced the word "Darwinism" for the theory, and minimized his own part in the elaboration of the theory as "one week to twenty years."

This sturdy band of converts assumed the defense of Darwin before the public, while Charles remained aloof. Sequestered in his estate at Down, he calmly answered letters and listened, but not too much, to the tumult over the horizon. "It is something unintelligible to me how anyone can argue in public like orators do," he confessed to Hooker. Hewett Watson, another botanist of note, wrote to him shortly after the publication of the *Origin:* "Your leading idea will assuredly become recognized as an established truth in science, i.e., 'Natural Selection.' It has the characteristics of all great natural truths, clarifying what was obscure, simplifying what was intricate, adding greatly to previous knowledge. You are the greatest revolutionist in natural history of this century, if not of all centuries."

Watson's statement was clairvoyant. Within ten years the *Origin* and its author were known all over the globe, and evolution had become the guiding motif in all biological studies.

Alfred Russel Wallace

I

If it had not been for the mere chance that Alfred Russel Wallace chose to dispatch the account of his discovery to Darwin, we might today be acclaiming him as the founder of modern biology.

Since the publication of Darwin's *Origin of Species,* innumerable addresses have extolled the scientific achievements of Darwin. Wallace, on the other hand, has been present in many of these historical accounts only as an attenuated shadow—a foil to the great Darwin. Many do not know his name, or, if they recognize it, do so only with the vague impression that Wallace found out something which Darwin had already perceived more clearly. The fact that Wallace emerged from a social class different from Darwin's, that he was not a product of the traditional schools, has perhaps militated subtly against the full recognition of his scientific achievements. Moreover, Wallace, by nature modest and retiring, never thrust himself forward and his contributions have in some cases passed into the body of scientific thought without acknowledgment.

Wallace was born on January 8, 1823, in the Welsh village of Usk in Monmouthshire. In this remote district of low rents and country food his parents had sought refuge from a series of financial misfortunes which had brought them to the verge of total poverty. His father appears to have been a well-intentioned but somewhat inept middle-class gentleman who by degrees had lost most of a small but comfortable inheritance. Although the family had thus fallen upon evil days before Wallace was born, it is worth noting that his father had had a good education, and was fond of books. As Wallace himself observes, "Through reading clubs or lending libraries we usually had some of the best books of travel or biography in the house." Moreover, his father was addicted to reading aloud to his wife and children in the evenings.

Before Wallace was fourteen, however, he had left home, and he saw little of his parents thereafter. Most of his formal schooling, which was scant, had been obtained in the grammar school of the old town of Hertford. He remembered into old age and with great affection the meadows beyond the town, and the millstream with its great, dripping waterwheel. In the law courts where the assizes were held he heard the trials of poor sheep-stealers who, in those days, might be liable to transportation for life. Such painful scenes were part of the harsh criminal code so vividly pictured in W. H. Hudson's *A Shepherd's Life.* Wallace never forgot these episodes. They are a strong element in his sympathy for the socially deprived and unfortunate which sets Wallace's thinking so apart from that of many of his Victorian colleagues. He never shared the enthusiasm of some of them for unrestricted struggle within the social world of man.

In 1837, after further family financial disasters, Wallace went to London, where he lived for a brief period with his older brother John. He was waiting for the return of another brother, William, from whom he was to learn surveying. Here he wandered about the marvelous streets of London, gazing into the newly installed plate-glass windows of the best shops, attending workingmen's lectures, and reading Thomas Paine's *Age of Reason.* Then, after a glimpse of this metropolitan fairyland, he went back into the country to become a surveyor. Here, a little like William Smith, the surveyor who had discovered the principles of stratigraphy only a few decades earlier, Wallace discovered the many fascinations of geology.

In the country around Barton, near the North Downs, between intervals of application to surveying, he wandered among streams and valleys bearing such appellations as Roaring Meg and the Devil's Dyke. He saw fossils lodged in ancient strata; he rambled in solitary fashion over a wild landscape of which he felt himself increasingly a part. In spite of Wallace's modest protestations in after years, one can observe the same deep sense of kinship with nature, the same prolonged and thoughtful observation, that had inspired Darwin.

Poor though Wallace was, accident had led him to an occupation which had enticed him into the open instead of confining him to a shop. It is not without historical interest that in the same period a great literary naturalist, Henry David Thoreau, had in America chosen a similar profession. Contemplating a mountain journey of his surveying years, Wallace once remarked: "We obtain an excellent illustration of how nature works in molding the earth's surface by a process so slow as to be almost imperceptible."

It was in this pursuit that Wallace spent the years that most young men of today give to high school and college. Nevertheless it was in many ways one of the happiest periods of the great naturalist's life. He was with a responsible older brother whom he loved and who cared for him with great tenderness. He was out in the open air, living a frugal but healthy existence. As he himself tells us, "If I had continued to be similarly employed after I became of age, I should most probably have become entirely absorbed in my profession. It seems unlikely that I should have ever undertaken a journey to the almost unknown forests of the Amazon."

In 1843 Wallace's father died; his brother, having no work in prospect, was forced to advise the youth that he must look out for himself. Once more Wallace drifted to London. He tells us that he was shy, frail and lacking in physical courage. Sensitive and a lover of solitude he undoubtedly was, but we must entertain reservations about his confession of cowardice. For this was the young man who, like the great English voyagers before him, was later to wander alone into the depths of the Amazon forest or venture by native *prau* among the dangerous islands of the

Malayan seas. As his friend James Marchant stated, "Wallace, at no time during these wanderings, had any escort or protection, having to rely entirely upon his own tact and patience, combined with firmness, in his dealing with the natives."

In 1844 Wallace obtained a small teaching post in a school at Leicester, where he stayed for a little over a year. Here in a good town library he continued his own reading interests, including Charles Lyell's *Principles of Geology,* Alexander von Humboldt's *Personal Travels in South America* and Thomas Malthus' *An Essay on the Principle of Population.* Years later, Joseph Hooker wrote to Darwin: "I think Humboldt is underrated nowadays. Well, these were our Gods, my friend, and I still worship at their shrines a little." In this wistful phrase Hooker has caught the essence of a bygone day. Humboldt, the great traveler, had had an influence on the youthful naturalists of Darwin's and Wallace's time which can only be compared to Linnaeus' great influence in the eighteenth century.

It can already be seen that, although they were separated by a vast social distance, Darwin and Wallace had pretty much the same reading background. Both had been stimulated by their scientific predecessors, Lyell and Humboldt. Darwin had already made his voyage and was working secretively at Down while Wallace, fourteen years younger, was still teaching children to read. No one could possibly have imagined that this unpromising, unlettered young schoolteacher would catch up to, and share honors with, the greatest biologist of the age, but so it was to be.

In the town library of Leicester, Wallace met and became fast friends with Henry Walter Bates, the entomologist with whom he was later to journey to South America. The death of William Wallace caused Alfred to resign his teaching position, and to move for a brief period to Neath. Wallace continued to correspond with Bates, however, and together they began to dream of a tropical collecting expedition.

The young men's letters reveal that both were greatly stimulated by Chambers's *Vestiges of the Natural History of Creation,* to which I have previously referred. "I begin to feel dissatisfied," Wallace wrote to Bates, "with a mere local collection. I should like to take some one family to study thoroughly, principally with a view to the theory of the origin of species."

These words were written twelve years before the *Origin of Species* was published. Wallace had read Darwin's *Journal of Researches;* he had also read the *Vestiges.* "I have rather a more favorable opinion of the *Vestiges* than you appear to have," Wallace wrote his friend Bates, and from then on his attention to the problem of evolution never wavered. Moreover, it is pleasing to record that, unlike some of the later Darwinists, he never abused Chambers. "The *Vestiges* is a book," he said in later years, "which has always been undervalued." He knew well that it had been a vital stimulus to many.

In the end, nevertheless, it was to be an American, William Henry Edwards, who was to center the plans of Bates and Wallace upon the Amazon as their first great adventure. In 1847 they chanced to read Edwards' book entitled *A Voyage up the River Amazon.* "The whole region north of the Amazon," wrote Edwards, himself a naturalist and collector of no mean descriptive powers, "is watered by numberless rivers, very many of which are still unexplored. It is a sort of bugbear country, where cannibal Indians and ferocious animals abound to the destruction of travelers. This . . . has always been Fancy's domain."

Scarcely had Wallace and Bates perused the book—with its accounts of rare animals, birds and insects, its observation of "monkeys who vary in species with every degree of latitude or longitude"—when the young naturalists hastened to the British Museum for advice. Here they were assured that the whole of northern Brazil was very little known and that collections made there would easily pay the expenses of the trip. By chance they discovered that Edwards was visiting London. Accordingly Bates and Wallace called upon him. He encouraged them in their ambitions and gave them letters to some of his friends in Pará. There was wide enthusiasm for natural history collections among the well-to-do in these years, and the young men were fortunate in securing a reliable agent to handle their collections and keep them informed of the market. Early in April of 1848 they sailed in the bark *Mischief* for Pará.

Their initiation to life at sea was remarkably like that of

Darwin's on the *Beagle.* Soon after they put to sea they were beset by a violent storm. A part of the bulwarks was carried away, and the ship almost foundered. For a week Wallace lay desperately ill from seasickness. Finally the storm passed and the ship had fine sailing weather for the rest of the voyage. Wallace was to remember into old age the exquisite blue of the tropic sea by day, and the shining phosphorescence of the ship's wake by night.

When they went ashore in Pará, he was awed and inspired, as Darwin had been, by the primeval forest. "Here," he wrote,

> no one who has any feeling of the magnificent and sublime can be disappointed; the somber shade, scarce illumined by a direct ray even of the tropical sun, the enormous size and height of the trees, most of which rise like huge columns a hundred feet or more without throwing out a single branch, the strange buttresses around the base of some, the spiny or furrowed stems of others, the curious and even extraordinary creepers and climbers which wind around them, hanging in long festoons from branch to branch, sometimes curling and twisting on the ground like great serpents, then mounting to the very tops of the trees, thence throwing down roots and fibers which hang waving in the air or twisting round each other, form ropes and cables of every variety of size and often of the most perfect regularity. These, and many other novel features—the parasitic plants growing on the trunks and branches, the wonderful variety of foliage, the strange fruits and seeds that lie rotting on the ground—taken altogether surpass description. . . . Here lurk the jaguar and the boa-constrictor, and here amidst the densest shade the bellbird tolls his peal.

This was to be Wallace's world for the next four years; a world where gecko lizards walked upside down on one's ceiling, where one's spoons, cups and bottles were made from gourds, where tree frogs were as gaily colored as children's toys, and where through the vast, illimitable forests the only thoroughfares were the enormous rivers emerging out of the unknown and hurrying with all manner of natural wreckage toward the sea. The spoonbill and the scarlet ibis stalked in the shallows. Overhead in the forest attic flew shrieking parrots. Innumerable monkeys, astonishing little caricatures of men, contemplated life while hanging by the ends of their prehensile tails from that same refuge. It was a world where the floods of the great rivers and the dimness of the forest had driven life to the upper levels of the branches.

As for man, whether primitive or civilized, the waterways were his only grasp upon reality, his only means of orientation, even his only means of emerging into the sun. For four years Wallace was to be a wandering corpuscle along the muddy arteries of a continent. At the end of that time, and after his young brother Herbert, who had bravely come out to join him in 1849, had died of yellow fever, Wallace decided to return home. "My health," he observed in his autobiography, "had suffered so much by a succession of fevers and dysentery that I did not consider it prudent to stay longer." Bates, his companion of the outward voyage, stayed for seven years more, producing in 1863 *The Naturalist on the River Amazon,* now a great English classic.

II

In 1852, Wallace stood in a leaking longboat of the brig *Helen* over a thousand miles from the port of Pará and watched almost everything he had collected or thought about through those four years in the depths of the Amazon jungle burning before him. Flames leaped from shroud to shroud of the abandoned vessel. Monkeys screamed and ran into the heart of the fire. Fire roared in his collections below decks, and ate at the masts until they fell. Only a single parrot, clinging to a burning rope on the bowsprit, dropped into the sea and was saved.

In the boats the men watched through the night, hoping that the pillar of flame would swing a passing ship out of her course to pick them up. The dawn came, charred and vacant. They were alone on the heaving sea with a limited supply of water and food.

Wallace watched while the captain set a course for Bermuda; he ate a biscuit numbly and helped the others bail the leaking boat. If he survived, he would return to England as poor as when he had journeyed out to South America four years before. It might have been better to have stayed and rotted with the other eccentrics in the fever ports of the tropics. Luck had been with him on the great river and in the camps of half-wild savages, but now, if there should be a storm, if they should be becalmed until the water ran out . . .

Ten days later, on short rations and still two hundred miles from Bermuda, the entire crew was picked up by a passing vessel bound for London. Wallace was now twenty-nine and there had been perhaps, only one gain: in his long, difficult traverses over the watershed of the world's greatest river, he had come to realize the enormous diversity of life, and how that diversity seemed related to geographic barriers often represented by the confluent fingers of the huge river. Even the fishes within the river differed. Those of the Amazon were peculiar to itself, but those of the upper tributaries were distinct. The number of separate species inhabiting the Amazon basin, he observed, "must be immense." He wondered a little about all that related diversity.

One other thing could be put down to experience. Alone among the great Victorian evolutionists, he had actually lived with primitive men. He had not just gazed at them politely from exploring vessels. He had ventured into the high Amazons; he had visited country untrod by Europeans. He had sweated with naked savages up dangerous river portages. He had drunk at their feasts, slept in their houses, observed every aspect of their lives. There was a touch of the anthropologist as well as the naturalist in Wallace. Later on in the century these anthropological interests would re-emerge and leave him somewhat isolated from his fellows. But now the Amazons had been left behind him. To what would he turn?

Coming ashore with nothing but the clothes he stood in, he sought refuge in the family of a married sister, and for the next eighteen months busied himself with a book on his travels. He also composed a now rare little book on palm trees, based on sketches which he had salvaged from the wreck. A small sum of insurance, through the foresight of his agent, helped him. "Fifty times," he wrote to a friend, "I vowed never to trust myself more on the ocean. But good resolutions soon fade and I am already only doubtful whether the Andes or the Philippines are to be the scene of my next wanderings." The tropics had got into his blood.

By now Wallace was reasonably well known to naturalists through his constant attendance at the meetings of the Zoological and Entomological societies in London. After much careful collecting of information from fellow travelers and scientists, he came to the conclusion that the Malay Archipelago offered excellent opportunities for an exploring naturalist who had to live by his collections. The islands were regarded as healthier than the Amazon, and Wallace determined to go there if he could secure free passage in some Government ship. Through the representations of Sir Roderick Murchison, then president of the Royal Geographical Society, this request was granted by the Government. The fact is worth noting because it indicates that the thirty-year-old author of two books was already favorably known in scientific circles before he left for the jungles of the East.

In April 1854 he had reached Singapore with a young English assistant. Behind them lay Suez. This was before the days of the canal, and Wallace in a letter gives a vivid picture of the journey across the neck of the isthmus in horse-drawn omnibuses along a desert road littered with the bones of camels. It was from Singapore, however, that, as Wallace himself wrote afterwards, "I was to begin the eight years of wandering throughout the Malay Archipelago, which constituted the central and controlling incident of my life."

"Singapore is rich in beetles," he exulted in a letter home, for beetles were a favorite collector's item. "I shall have a beautiful collection of them." The assistant, Charles, he confides, "is doing the flies, wasps and bugs. I do not trust him yet with beetles."

In Malacca he suffered again from fever. "I went to the celebrated Mount Ophir and ascended to the top, sleeping under a rock. The walk there was hard work, thirty miles through jungle in a succession of mud-holes, and swarming with leeches which crawled all over us and sucked when and where they pleased. . . . I got some fine new butterflies there and hundreds of other new or rare insects. Huge centipedes and scorpions, some nearly a foot long, were common."

He speaks in turn and with equal interest and affection of tigers and pitcher plants, orangutans and birds of paradise. "The more I see of uncivilized people," he wrote in a fashion of which more would be heard later, "the better I think of human nature on the whole, and the essential differences between civilized and savage man seem to disappear."

Over the map of Indonesia the spidery lines of his travels lengthened year by year. Marchant, his editor and biographer, records that during the eight years of his eastern sojourn he traveled some 14,000 miles within the archipelago and collected well over 125,000 specimens. He was the first Englishman to see birds of paradise in their natural habitat. He almost fainted with excitement and esthetic delight when he captured one of the rare bird-winged butterflies of iridescent green.

Yet if this last should suggest to some minds an effeminate character, let this record of a single journey in a Malay sailing vessel suffice to reveal the iron in his character:

> My first crew ran away; two men were lost for a month on a desert island; we were ten times aground on coral reefs; we lost four anchors; our sails were devoured by rats, the small boat was lost astern; we were thirty-eight days on the voyage home which should have taken twelve; we were many times short of food and water; we had no compass lamp owing to there not being a drop of oil in Waigiou when we left; and to crown it all, during the whole of our voyage, occupying in all seventy-eight days, we had not one single day of fair wind.

This voyage was but one of the many of which he writes about "laying in stores, hiring men, paying or refusing to pay their debts, running after them when they try to run away, going to the town with lists of articles *absolutely necessary* for the voyage, and finding that none of them could be had for love or money, conceiving impossible substitutes and not being able to get them either." Nevertheless he can turn from these tribulations to say to his brother-in-law: "Your ingenious arguments to persuade me to come home are quite unconvincing. I have much to do yet. I am engaged in a wider and more general study—that of the relations of animals to space and time." During this period of deep thought he also wrote to a boyhood friend, George Silk: "You cannot imagine how I have come to love solitude. I seldom have a visitor but I wish him away in an hour."

Although Wallace appears to have been a believer in the general principle of evolution since the time of the *Vestiges,* it was here in Malaya, amidst fever and the long indoor days of the rainy season, that he began to marshal the thoughts that would place him beside Darwin. Wallace was tremendously interested in plant and animal distribution. He studied William Swainson and Humboldt. He now had had experience in both the New World and Old World tropics. The question of the origin of species, he tells us, "was rarely absent from my thoughts." Traveler though he was, he carried books with him. "The great work of Lyell," he remarks in his autobiography, "had furnished me with the main features of the succession of species in time." By combining Lyell's observations with the facts of animal distribution, Wallace thought that "valuable conclusions might be reached." In thus using data from more than one field he was revealing a synthesizing power not unlike Darwin's.

While visiting Borneo in 1854 and 1855 he prepared a paper, "On the Law Which Has Regulated the Introduction of New Species." He sent it to the *Annals and Magazine of Natural History,* where it was published in September 1855. "Every species," Wallace wrote as his principal conclusion, "has come into existence coincident both in space and time with a pre-existing closely allied species." This principle is not, of course, Darwinian natural selection, but it points to the fact that only organic change in time could explain such a close connection between the life of the present and that of the past. It was another way of driving home the fact that biological evolution was the only rational explanation for linking living forms with related extinct fossils.

Wallace was not, of course, quite as original as he thought when he wrote this paper; the idea has a long history running back through Darwin to John Hunter. But there was no way for the young Wallace, far from the great libraries, to realize this. Furthermore, the older statements of the principle are less precise. Wallace had a gift for clear, incisive statement, and it emerges in this, his first evolutionary paper. The young naturalist was disappointed when the article passed almost unnoticed. Darwin, with whom he had now had some correspondence, comforted him. Lyell and Edward Blyth, the student of the South Asian faunas, Darwin wrote, had both called it to his attention.

Once more Wallace turned to the practicalities of the sea and the islands, setting forth on a search for birds of paradise. There is evidence, however, that a rarer, wilder search was by then running in the huntsman's head. In 1858 he referred once more to his paper. "It is," he wrote to his old friend Bates, "merely the announcement of the theory, not its development." He goes on: "I have been much gratified by a letter from Darwin. He is now preparing his great work on 'Species and Varieties,' for which he has been collecting materials twenty years." The letter indicates that he had no notion of Darwin's theory, though he looked forward to its publication with curiosity. Just two months later the huntsman had stumbled upon his quarry. In the mind of the solitary wanderer the present had become one with the past.

III

During the early months of 1858 Wallace was living at Ternate in the Molucca Islands off the western tip of New Guinea. He was suffering severely from intermittent fever, and during one of these attacks, while he lay weak but lucid on his bed, his mind began to revolve upon the "species problem" which had fascinated him since those days when he had read the *Vestiges.* "Something," he says, "brought to my recollection Malthus's *Principle of Population* which I had read about twelve years before."

In a lightning flash of insight, it occurred to the feverish naturalist that Malthus' checks to human increase—accident, disease, war, and famine—must, in similar or analogous ways, operate in the natural world as well. "Vaguely thinking over the enormous and constant destruction which this implied, it occurred to me," he tells us, "to ask the question, 'Why do some die and some live?' " The answer, Wallace felt, was clear: the best fitted live. "From the effects of disease the most healthy escaped; from the enemies the strongest, the swiftest, or the most cunning; from famine, the best hunters."

Again his mind leaped forward. "Considering the amount of individual variation that my experience as a collector had shown me to exist, then it followed that all the changes necessary for the adaptation of the species to the changing conditions would be brought about; and as great changes in the environment are always slow, there would be ample time for the change to be effected by the survival of the best fitted in every generation.

"I waited anxiously for the termination of my fit," Wallace goes on with unconscious humor, "so that I might at once make notes for a paper." In two evenings he was ready to dispatch it to Darwin. "I hoped," he said, "that the idea would be as new to him as it was to me."

The paper reached Darwin at Down in June 1858. Wallace's innocent elation was to be Darwin's despair. "All my originality, whatever it may amount to, will be smashed," Darwin wrote to Lyell on the same day. "I never saw a more striking coincidence. . . . Your words [here Darwin was referring to Lyell's earlier warnings that he might be anticipated] have come true with a vengeance."

In a genuine agony of spirit Darwin sought the advice of his friends. Could he now publish honorably? "I would far rather burn my whole book," he protested, "than that he or any other man should think that I had behaved in a paltry spirit."

The friends' counsels prevailed. Darwin, as it happened, had a copy of a letter sent to Asa Gray at Harvard College, which validated his priority, and it was decided to read Wallace's paper along with Darwin's letter to Gray and an extract from his unpublished sketch of 1844, before a meeting of the Linnean Society on July 1, 1858. This was the dramatic prelude to the great intellectual storm which would shake the latter half of the nineteenth century and be reflected in the Scopes trial of the 1920s in the United States.

Yet the beginning was deceptively quiet. Wallace was traveling among the islands, unaware of fame, lost in a dream of rare birds and rarer butterflies. Darwin, who was nursing a sick child, did not attend the meeting. There was no discussion before the Society. "The interest," Hooker commented later, "was intense, but the subject was novel."

The Fellows of the Society were overawed by the tacit approval of such prominent scientists as Hooker and Lyell. The storm, the controversies, would come later with the publication of the *Origin of Species* in the following fall of 1859. Wearily Darwin picked up his pen once more, but finally, in better humor, he wrote to Hooker, "I find it amusing and improving work. I am now most heartily obliged to you and Lyell for having set me on this. I confess I hated the thought of the job."

A man pursuing birds of paradise in a remote jungle did not yet know that he had forced the world's most reluctant author to disgorge his hoarded volume, or that the whole of Western thought was about to be swung into a new channel because a man in a fever had felt a moment of strange radiance. The path led on. It always did for Wallace. In 1862, aware now, in some degree, of the change in his fortunes, he turned homeward. Two roach-eating birds of paradise were his companions. Wallace has unwittingly left us a vivid picture of sea travel in his time. "Every evening," he explains, "I went to the storeroom in the fore part of the ship where I was allowed to brush the cockroaches into a biscuit tin." At Malta a bakery supplied the roaches. The birds and Wallace arrived home in good health.

IV

It is comparatively easy to give a fair account of a man who dies within his own generation. He is encapsulated in his own time, and his follies and achievements can be comprehended accordingly. It is not so simple to evaluate the career of the man for whom Wallace's Line, dividing the faunal region of continental Asia from that of New Guinea, is named. He belonged to no cliques, and avoided participation in virulent scientific and theological disputes. In terms of his later life he can be classed with

Asa Gray in America as one of the more theistically inclined evolutionists who paved the way for the acceptance of evolution in the later years of the century. He was in some respects a more able anthropologist than Darwin, and it is significant that when the two men came to differ—and differ they eventually did—it was over man that their disagreement arose.

In 1864 Wallace composed a paper which led eventually to strong intellectual differences, but never a personal break, with Darwin. In this paper he gave vent to a new view: namely, that with the rise of the human brain a creature had emerged who, for the first time in the long history of life, had escaped from the specialization of *parts* toward which evolution seemed always to progress.

With man this process was apparently at an end. Man, in his brain, had developed a specialized organ whose whole purpose was to enable him to escape specialization. He could now increasingly assign to his clothing and implements the special activities for which the animal had to develop organs in its own body. Man, by contrast, could take off and put on. Specialization could be left to his cultural shell, his technology. Armored within that shell, great-brained man was in the process of acquiring a sort of timeless, unchanging body in the midst of faunas and floras still forever evolving and vanishing.

Wallace did not deny that small alterations might still be taking place in man, but he regarded them as insignificant. "With our advent," Wallace maintained, "there had come into existence a being in whom that subtle force we term *mind* became of far more importance than mere bodily structure." We have taken away from nature the power of change which she exercises over all other animals. "Man does this by means of his intellect alone," Wallace argued, "which enables him with an unchanged body still to keep in harmony with the changing universe." Wallace, leaping beyond his colleagues, had glimpsed the full anthropological significance of the evolved brain. The idea was hailed by prominent thinkers. John Fiske spoke of it as opening up "an entirely new world of speculation."

Darwin expressed his admiration of the paper to Hooker, calling it "most striking and original." He agreed to the leading idea, and expressed the wish that Wallace had written Lyell's chapters on man. It was only later, as Wallace's bepuzzlement over man increased, that the two men came to a parting of the ways.

"Natural selection," pondered Wallace, "could only have endowed savage man with a brain a few degrees superior to that of an ape, whereas he actually possesses one very little inferior to that of a philosopher." Man's curious hairlessness, the structure of the human larynx, and other odd human features began to loom impressively in Wallace's thinking. Finally the man who had not been impressed in his youth by organized religion was led to suggest that a higher intelligence might have played a hand in the development of our kind.

"I differ grievously from you, and am very sorry for it," wrote Darwin courteously.

Huxley was severely critical.

Hooker wrote to Darwin of Huxley's remarks: "The tumbling over of Wallace is a . . . service to science."

This was curious emotionalism concerning a man who did much to destroy narrow Victorian racial prejudices, and whose contributions on a variety of subjects were of such lasting importance.

Wallace lived to be almost ninety-one. As a young man he had surveyed for the first railroad lines in England. He had been shipwrecked in the days of sail. Yet it was his fate to live on into the days of the airplane and the automobile. In a world where human generations are short, this means that Wallace lived on into a world increasingly alien and strange to him. In some ways he found himself old-fashioned; in some ways, percipient of the future. He ranged farther in his mind and in his interests than his more contained and scientifically orthodox colleagues. Sometimes this habit led to highly successful insights, sometimes to disastrous failures, but through all his vicissitudes one is conscious of being in the presence of a fertile and inquiring mind.

His was a full life—that much can be said. A lengthy biographical study would have to incorporate much that cannot be included here: his growing interest in spiritualism, his odd antipathy to vaccination, his lifelong defense of the poor and helpless, his bitter memories of the worst of the Industrial Revolution. Many of his ideas were mistaken. Some forecast the future with uncanny perception. Some, as we have seen, outran with a flashing brilliance even his friend and colleague, Darwin.

Wallace never compromised his beliefs even when they caused contemptuous comment among his associates. He was an individualist of a sort rarely seen in the "team science" of the modern era.

"If you would learn the secrets of nature," Henry David Thoreau once remarked, "you must practice more humanity than others." It is the voice of a contemporary in the years before the green continents were despoiled. Wallace would have understood the words.

Charles Lyell

I

"I FEEL AS IF my books," Charles Darwin once confessed, "came half out of Sir Charles Lyell's brain." The great biologist was admitting to no more than the simple truth. Sir Charles Lyell, who remained until late in his career a reluctant evolutionist, was paradoxically the ground-breaker for the triumph of the *Origin of Species.*

Lyell is remembered chiefly as a founder of modern historical geology. But he was also a biologist whose studies form the backbone of the achievement of both Darwin and Wallace. In his day he addressed tabernacles in both England and America full of people eager to hear the world-shaking views of the new geology. Today the man in the street has forgotten him. By a curious twist of history, Darwin replaced Lyell as a popular idol. Yet this gaunt-faced man who ended his days in near-blindness was one of the greatest scientists in a century of distinguished men.

A generation before Darwin he took a world of cataclysms, supernatural violence and mystery, and made of it something plain, expected and natural. It was as though we had been unable to see the earth until we observed it through the eyes of Lyell. Ralph Waldo Emerson wrote at mid-century: "Geology, a science of forty or fifty summers, has had the effect to throw an air of novelty and mushroom speed over entire history. The oldest empires—what we called venerable antiquity—now that we have true measures of duration, show like creations of yesterday. . . . The old six thousand years of chronology becomes a kitchen clock." Geologic time is now so commonplace that the public forgets it once had to be fought for with something of the vigor that was later to be transferred to the evolutionary debates of the 1860s.

Today, as the splitting-up of science into numerous special disciplines leaves Lyell one of the founders of historical geology, the world has tended to forget that he also wrote extensively upon zoological subjects, or that, in the words of the great U. S. evolutionist Asa Gray, "Lyell is as much the father of the new mode of thought which now prevails as is Darwin."

Yet in the first years of evolutionary controversy, beginning with Jean Baptiste de Lamarck, Lyell found himself popularly arrayed with the resistance to evolution. He was not to alter his public position until the autumn of his life. To us it may seem an almost willful rejection of the new age of science. Oddly enough we are wrong. Reading history backward is almost worse than not reading history at all. One must live both in a given time and beyond it to appreciate at once its complexities and its half-veiled insights.

Lyell's rejection of evolution was one of the first rational products of the new geology. A hint as to the nature of the situation is to be found in a passage in Lyell's *The Antiquity of Man,* published in 1863. The issue, to our modern eyes, is obscured by the terms in which it was argued. Lyell wrote:

> It may be thought almost paradoxical that writers who are most in favor of transmutation (Mr. C. Darwin and Dr. J. Hooker, for example) are nevertheless among those who are most cautious, and one would say timid, in their mode of espousing the doctrine of progression; while on the other hand, the most zealous advocates of progress are, oftener than not, very vehement opponents of transmutation. We might have expected a contrary leaning on the part of both.

It is in the words "transmutation" and "progression," now unfamiliar, that the key to this mystery lies. When we come to know their significance, we will have learned that the road to the acceptance of evolution had unexpected turnings which, as we look backward, seem to have vanished, but which were real enough to the men of the nineteenth century. Before we can understand Lyell's position, this queer order of events has to be explored and comprehended.

II

Charles Lyell was born, the first of ten children, to well-to-do parents in Scotland in 1797. His father possessed a strong interest in natural history and may have helped unconsciously to guide his son's interests in that direction. As Darwin and Wallace were later to do, the young Lyell collected insects in his boyhood. Absent-minded but versatile, tree-climber and chess player, he matriculated at Exeter College, Oxford, in 1816. Having early stumbled upon a copy of Robert Bakewell's *Introduction to Geology* in his father's library, he sought out Dean Buckland's geological lectures at Oxford, and from then on was a haunter of chalk pits, rock quarries, caves and river terraces.

In 1818 he made the usual continental tour with his parents and sisters. The slow carriage travel of that day promoted leisurely observation, and Charles made the most of it. He saw the red snow and glaciers of the high Alps as well as the treasures that lie open to the observant in the flints of the common road. Lyell had not as yet settled upon a career in geology. He was destined for the law, and shortly after his graduation from Oxford he came to London to prepare himself for the bar.

Even in London, however, Lyell was soon elected a Fellow of the Geological Society and joined the Linnean Society. Two handicaps tended to retard his legal career. His eyes were weak and troublesome, and he suffered from a slight speech difficulty, with which he was to contend bravely in his years as a lecturer on the natural sciences. When he was called to the bar in 1825 he was already contributing articles on scientific subjects to the *Quarterly Review.*

It has sometimes been intimated that Lyell was "only an armchair geologist," that he was scientifically timid, a rich man's son who happened to dabble his way to success in a new science. But in those days there was little in the way of public support for science. Even the great schools were still largely concentrated upon the classical education of gentlemen. Only the man of independent means, like Lyell or Darwin, could afford to indulge his interest in science. With occasional struggling exceptions such as Wallace, it was the amateur who laid the foundations of the science of today. The whole philosophy of modern biology was established by such a "dabbler" as Charles Darwin, who never at any time held a professional position in the field. Charles Lyell and his great precursor, the Scotsman James Hutton, similarly laid the foundations of modern geology without claiming much in the way of formal institutional connections. Important though institutional and government support has come to be, it has led to a certain latent snobbery in professional circles. The amateur has had his day. But his was the sunrise of science, and it was a sunrise it ill becomes us to forget.

The charge that Lyell was an armchair worker will not stand against the facts. But even if the accusation held, the whole question would turn on what came out of the armchair. In actuality Lyell in his younger years made numerous trips to the continent to examine the evidence of geology at first hand. Later, in the 1840s, he visited America, where he made similar ventures into the field, even though he was by then lecturing to thousands. As his biographer T. G. Bonney remarks, "Whenever there was hope of securing any geological information or of seeing some remarkable aspect of nature, Lyell was almost insensible either to heat or to fatigue." It is hard to see how a man suffering from bad eyes could have done more.

In support of the charge of scientific timidity, it is observed that Lyell opposed himself for many years to evolutionary views; it is said that his public and his private statements upon this score were vacillating. "How could Sir Charles Lyell," wrote one of Darwin's contemporaries, "for thirty years read, write and think on the subject of species and their succession, and yet constantly look down the wrong road?" From the vantage point of over a century this question can be answered. A whole new theory of life and time is not built by one man, however able. It is the product of multitudinous minds, and as a consequence it is also compounded of the compromises and hesitations of those same minds. Later, when the new world view comes to be ascribed, as it generally is, to a single individual such as Darwin the precursors of the discoverer begin to seem incredibly slow-witted.

Whatever men may think on this score, however, the record shows that Lyell was a man of intellectual courage. He entered the geological domain when it was a weird, half-lit landscape of gigantic convulsions, floods and supernatural creations and extinctions of life. Distinguished men had lent the power of their names to these theological fantasies. Of the young Lyell, the "timid" Lyell who later strained Darwin's patience, a contemporary geologist wrote: "He stood up as a reformer, a radical reformer, denouncing all the old notions about paroxysms and solving every geological question by reference to the action of constant and existing physical causes. Never had a revolutionist harder work to get a sober hearing, or less prospect of overturning the works and conclusions of other men."

III

Geology at the beginning of the nineteenth century was known to many in England as a dangerous science. As such it both attracted and repelled the public. A body of fact and interpretation had arisen that could only be kept in accord with the Scriptural interpretation of earth's history by the exercise of considerable ingenuity. Theological authority was strong, and there was the greatest pressure upon geologists to avoid direct conflict with the church. Moreover, some of the early geologists were primarily theologians themselves, and were understandably anxious to reconcile geology with their religious beliefs. By degrees there had thus arisen a widely accepted view of geological history known as catastrophism.

This orthodox geological creed was an uneasy amalgam of the new scientific facts seen in the flickering, unreal light of mythological and romantic fantasy. Unlike the slow evolutionary successions that we recognize today, the record of geology was held to contain sudden catastrophic breaks. Mountain ranges were thought to be heaved up overnight; gigantic tidal waves, floods, paroxysms of the earth's crust were thought to mark the end of periods of calm. At such hours life vanished only to be restored through renewed divine creation, taking in the new period a more advanced form, and pointing steadily on toward the eventual emergence of man. It may thus be observed that the students of catastrophism had become aware of organic change in the rocks, but they saw the planet as having been molded by forces seemingly more powerful than those at work in the present day, and thus by implication supernatural.

Awareness of a succession of life-forms in the strata of the earth had been slowly increasing since the close of the eighteenth century. Furthermore, it was seen that these extinct forms of life showed an increasing complexity as one approached the present. Since the record of the land vertebrates is particularly incomplete and broken, there arose the idea that, instead of a genuine continuity of life from age to age, the breaks in the geological record were real breaks. There had been a genuine interruption between the life of one age and that of another; each geological period had its own flora and fauna largely distinct from that which preceded and followed it. The slow, grand progression of life was seen as through a jerky, discontinuous, ill-run motion picture.

Men still understood neither the real age of the earth nor the fact that the breaks they found in the records of the rocks were not world-wide, but rather only local discontinuities. The imperfections of the geological record, or the passages between the discontinuities, could only be learned through the piling up of empirical evidence, a task that had only begun.

The catastrophic school had a powerful religious appeal. It retained both the creative excess and fury of an Old Testament Jehovah. "At succeeding periods," wrote Adam Sedgwick, one of Darwin's geological teachers at Cambridge,

> new tribes of beings were called into existence, not merely as the progeny of those that had appeared before them, but as new and living proofs of creative interference; and though formed on the

> same plan, and bearing the same marks of wise contrivance, oftentimes unlike those creatures which preceded them, as if they had been matured in a different portion of the universe and cast upon the earth by the collision of another planet.

People thought in terms of a geotheological drama, a prologue to the emergence of man on the planet, after which no further organic developments were contemplated. The theory predicted a finished world which, in some eyes at least, could be compressed into the figurative week of the Book of Genesis. "Never," commented Lyell, "was there a dogma more calculated to foster indolence, and to blunt the keen edge of curiosity, than this assumption of the discordance between the former and the existing causes of change."

In this half-supernatural atmosphere, Lyell in 1830 published the first volume of his *Principles of Geology*. Like most great ideas, its thesis was not totally original with the author. But to Sir Charles belongs the unquestioned credit of documenting a then unpalatable truth so effectively and formidably that it could no longer be ignored. In this respect again his career supplies a surprising parallel to that of Darwin. For Darwin too, at a later time, was the resurrector and documenter of forgotten and ill-used truths.

Lyell's principal precursor, James Hutton, died in intellectual eclipse in 1797, the very year that saw the birth of the man who was to revive his views—so tenuous and yet so persistent is the slow growth of scientific ideas. In the 1780s Hutton made the first organized and comprehensive attempt to demonstrate that the forces that had shaped the planet—its mountains, boulders and continents—are the same forces that can be observed in action around us today. Hutton had an ear for the work of raindrops, an eye for frost crystals splitting stones, a feel for the leaf fall of innumerable autumns. Wind and frost and running water, given time enough, ran his argument, can erode continents. Peering into the depths of the past, he could see "no vestige of a beginning, no prospect of an end."

Hutton, though not the first to suspect the earth's antiquity, and the work that perfectly natural forces can perform, was the first to write learnedly and extensively upon the subject. His work fell, however, into undeserved neglect. He was criticized as irreligious. In England the catastrophism theory, with its grander scenery and stage effects, had a more popular appeal. The world of Hutton by contrast was an unfinished world still unrolling into an indeterminate future. Its time depths were immeasurable, and the public had recoiled from its first glimpses into that abyss.

Yet this was the domain, and this the philosophy, upon which Lyell was to force his colleagues to take a long second look. He was a more eloquent and able writer than Hutton. But beyond this he had the advantage of almost fifty years of additional data, including his own personal study of the continental deposits. "Lyell," remarks one of his contemporaries, "was deficient of power in oral discourse, and was opposed by men who were his equals in knowledge, his superiors in the free delivery of their opinions. But in resolute combats, yielding not an inch to his adversaries, he slowly advanced upon the ground they abandoned, and became a conqueror without ever being acknowledged as a leader."

By degrees the idea of gradual change (uniformitarianism, as it came to be called) succeeded the picture of world-wide catastrophes. Supposition and quasi-theological imaginings gave place to a recognition of the work of natural forces still active and available for study in the world about us. The disjoined periods of the catastrophists began to be seen as one continuous world extending into a past of awe-inspiring dimensions. The uniformitarian school began to dominate the geological horizon. With the success of the *Principles* Lyell became one of two or three leading figures in English natural science. It is no wonder that the young Darwin, just returned from the voyage of the *Beagle,* gravitated so quickly to Lyell, whose revision of geology was to make his own triumph possible.

IV

Sir Charles Lyell had been raised in a more orthodox home than Charles Darwin. In fact, he was to confess in afteryears that it cost him a severe struggle to renounce his old beliefs. Nevertheless, in reviving the conception of limitless time, and in abandoning the notion of world-wide breaks in the geological record as urged by the catastrophists, Lyell was inevitably forced to confront the problem of life itself in all its varied appearances. His great predecessor, Hutton, had been largely able to avoid the issue because of the lack of paleontological information. In Lyell's time, however, the questions pressed for answer.

The catastrophist doctrine had given birth to a kind of romantic evolutionism to explain the increasing complexity of life. This was the doctrine of "progression" which Lyell opposed in many of his writings from the time of the *Principles* onward. Progressionism was the product of the new paleontology which had discovered differences among the life-forms of successive geological eras. The theory can be said to have borrowed from Lamarck the conception of a necessary advance in the complexity of life as we ascend through the geological strata to the present. Instead of establishing biological continuity (the actual physical relationship between one set of forms and their descendants) the progressionists sought to show only a continuity or an organic plan in the mind of God between one age and another. There was, in other words, no phylogenetic relationship on the material plane between the animals of one era and those of a succeeding one.

Progressionism thus implied a kind of miraculous spiritual evolution which ceases only when the human level has been attained. The idea is confusing to the modern thinker because he tends to read back into this literature true evolutionary connotations that frequently were not intended by these early writers. The doctrine is interesting as a sign of the compromises being sought between an advancing science and a still powerful religious orthodoxy.

"I shall adopt a different course," the young Lyell had written when he was contending for the uniformitarian view in geology. "We are not authorized in the infancy of our science to recur to extraordinary agents." The same point of view led him, in company with T. H. Huxley, Joseph Hooker and, later, Darwin, to reject the claims of progressionism. All of these men,

Lyell foremost among them, were uniformitarians in geology. They believed in the play of purely natural forces upon the earth. They refused or were reluctant to accept the notion of divine interposition of creative power at various stages of the geological record. They felt in their bones that there must be a natural explanation for organic as well as geological change, but the method was not easily to be had. Since Lyell was the immediate parent of the new geology, and since he was committed to natural processes, he was continually embarrassed by those who said: "You cannot show how nor why life has altered. Why then should we not believe that geological changes are equally the product of mysterious and unknown forces?"

We are now at the crux of the reason why Lyell was dubious about notions of "transmutation"—the term then reserved for ideas implying true physical connection among the successions of species or, as we would say, "evolution" from one form to another. Lyell's attitude toward evolution was influenced by the antipathy that he felt toward progression, toward the unexplainable. In bracketing the two together he in effect was indicating the need of a scientific explanation of organic change, if change indeed was demonstrable.

Beginning with the *Principles of Geology,* in which the second volume and part of the third are devoted to biological matters, Lyell had sought to examine the biological realm with an eye to answering the challenge of the catastrophist progressionists. As a consequence he came close to, but missed the significance of, the natural-selection hypothesis which was to establish the fame of Darwin. It was here that he took the wrong turning that led him away from evolution. Yet ironically enough, though Lyell failed to comprehend the creative importance of natural selection, he did not miss its existence. In fact, through a strange series of circumstances discovered in the literature, it is likely that he was fundamentally instrumental in presenting Darwin with the key to the new biology. He was so concerned, however, to array the evidence against the doctrine of progression that he missed the support that the same evidence gave for a rational explanation of the origin of species.

Against the progressionists' idea of mass extinction at each break in the geological record he cited the imperfections in that record. "There must," he contended, "be a perpetual dying out of animals and plants, not suddenly and by whole groups at once, but one after another." Although he had not solved the problem of the emergence of new forms of life, Lyell by arguing for geological continuity was bringing the question of extinction and of the origin of new species within the domain of scientific investigation.

He countered the progressionist hypothesis with a short-lived "nonprogressionism" in which he argued that the discovery of higher forms of life in older strata would demonstrate that the progressionist doctrine was based solely upon the fallible geological record. This retreat from straight evolution on the part of Lyell was somewhat wavering, but it continued into the 1850s. There is no doubt that it was an attempt philosophically to evade a problem which threatened to interpose into his system something miraculous and unexplainable that savored of the catastrophist doctrines he had struggled for so long to defeat. Only with the triumph of Darwin would a uniformitarian, a "naturalistic" explanation for the mutability of life be available to the uniformitarian followers of Lyell. It was only then that Huxley, Hooker and finally Lyell himself became converts to evolution, at a time when it was still being resisted by such men as Sir Richard Owen and Louis Agassiz—old-style catastrophists and progressionists who at first glance one might think would have eagerly embraced the new doctrine of genuine physical evolution.

Although the question has been obscured by hazy terminology, Lyell had already described before Darwin the struggle for existence and, up to a certain point, natural selection. He had not, however, visualized its creative aspect. Lyell made the first systematic attempt to treat the factors affecting the extinction of species and the effects of climatic change upon animal life throughout the long course of ages. "Every species," Lyell contended, "which has spread itself from a small point over a wide area must have marked its progress by the diminution or the entire extirpation of some other, and must maintain its ground by a successful struggle against the encroachments of other plants and animals." He goes on to speak of "the tendency of population to increase in a limited district beyond the means of subsistence." Nor was Lyell unaware of plant and animal variation, although he believed such variation to be limited. "The best-authenticated examples of the extent to which species can be made to vary may be looked for in the history of domesticated animals and cultivated plants," he wrote, long before Darwin's investigations.

But Lyell did more than this. In the *Principles of Geology* he marshaled a powerful attack on the possibility that new evolutionary forms might be able thus to maintain or perfect themselves. Lyell advanced what he called his principle of "preoccupancy." In essence this principle simply assumes that creatures or plants already well fitted for occupying a given ecological zone will keep any other forms from establishing themselves in the new habitat, even assuming that the competitors are capable of evolving. "It is idle," said Lyell, "to dispute about the abstract possibility of the conversion of one species into another, when there are known causes so much more active in their nature which must always intervene and prevent the actual accomplishment of such conversions." Using a number of contemporary examples Lyell sought to show that local alterations, say that from marsh to dry land, or fresh to brackish water, would never permit of slow organic change, because long before the organisms of the older environment could alter they would die out in competition with already adapted forms intruding to take advantage of the new conditions.

Lyell, in other words, could see how time and changing conditions might alter the percentages of living forms in given localities or change the whole nature of a flora. He understood that "the successive destruction of species must now be part of the regular and constant order of nature." What he still failed to grasp was that he was observing the cutting edge of the natural-selection process in terms of its normal, short-time effects. The struggle in nature that had so impressed him he had seen, if anything, too vividly. There was left no refuge, no nursing ground, by which the new could come into existence. The already created, the already fit, dominated every niche and corner of the living world. Lyell understood ecology before Darwin. He saw the web of life, but he saw it so tightly drawn that nothing new could emerge from it. As geographical or climatic conditions altered in the course of geological time, already existing forms moved from one area to another; he could see no evidence for a mechanism to explain the emergence of new forms.

His vision of the history of life was not wrong; it was simply incomplete. Lyell himself realized the complexities of the problem that beset him: there was a going-out without an equivalent coming-in, an attrition without a compensating creation.

"The reader will immediately perceive," Lyell wrote, "that amidst the vicissitudes of the earth's surface, species cannot be immortal, but must perish one after the other, like the individuals which compose them. There is no possibility of escaping this conclusion without resorting to some hypothesis as violent as that of Lamarck." Drawing back from this gulf, Lyell returns again and again to nonprogressionism. Nevertheless, like many naturalists of his day, he was willing to recognize "a capacity in all species to accommodate themselves, to a certain extent, to a change of external circumstances, this extent varying greatly, according to the species." Lyell recognized minor varietal differences of a seemingly genetic character in some animals. Beyond this he did not venture.

As one surveys the long record of his life, as one sees his influence upon Darwin and Wallace, as well as upon many other aspiring workers, one comes to recognize that to a major degree he set the scientific tone of the Victorian age. He brought to bear upon scientific thought and speculation a mind trained to the value of legal evidence. He was, on the whole, dispassionate, clear-headed and objective. By precept and example he transmitted that heritage to Darwin. He emphasized synthesis and logical generalization from facts. Both men eschewed small works and both amassed great bodies of material to carry their points. Lyell warned Darwin away from petty scientific bickering as a waste of time and nerves. At almost every step of Darwin's youthful career Lyell was an indefatigable guide and counselor. Then at that critical hour when Darwin learned of Wallace's independent discovery of natural selection, it was Lyell and Hooker who counseled the simultaneous publication of the papers of both men.

Darwin and Wallace were Lyell's intellectual children. Both would have failed to be what they were without the *Principles of Geology* to guide them. In science there is no such thing as total independence from one's forerunners. It is an illusion we sometimes like to foster, but it does not bear close examination. Even our boasted discoveries are often in reality a construct of many minds. We are fortunate if we sometimes succeed in fitting the last brick into such an edifice. Lyell himself knew this and tasted its irony.

He died in 1875 at the end of a long, outwardly uneventful life spent largely in the company of a beautiful, gracious and intelligent woman. After his wife's death in 1873 the light began to pass away from Lyell; he did not long survive her. A few years before, he had written to Ernst Haeckel: "Most of the zoologists forget that anything was written between the time of Lamarck and the publication of our friend's *Origin of Species.*"

Much indeed had been forgotten. In this little sigh of regret Lyell was even then resigning his hold upon the public which had once idolized him. To the true historian of science, however, he remains the kingmaker whose giant progeny, whether acknowledging their master by direct word or through the lines of their books, continue today to influence those who have never heard his name.

Of Charles Lyell, Darwin said what is so often remarked in our day of Darwin himself: "The great merit of the *Principles* was that it altered the whole tone of one's mind and, therefore, that when seeing a thing never seen by Lyell, one yet saw it partially through his eyes."

Charles Darwin, Edward Blyth, and the Theory of Natural Selection

I

WHETHER HIS VISIT to the Galápagos was the single event that mainly led Darwin to the central conceptions of his evolutionary mechanism—hereditary change within the organism coupled with external selective factors which might cause plants and animals a few miles apart in the same climate to diverge—is a point upon which Darwin himself in later years shed no clear light.

In his autobiography he says of the early period after his return from the voyage: "Nor did I ever intermit collecting facts bearing on the origin of species; and I could sometimes do this when I could do nothing else from illness." Finally, by Darwin's own account, he chanced to read the political philosopher Thomas Malthus in October 1838. He claimed to have seen, in the latter's treatment of the struggle for existence among human beings, the key to natural selection in the animal world.

> Being well prepared to appreciate the struggle for existence which everywhere goes on [he said], it at once struck me that under these circumstances favorable variations would tend to be preserved and unfavorable ones to be destroyed. The result of this would be the

formation of new species. Here then, I had at last got a theory by which to work. . . .[1]

This account of the Darwinian discovery has been hallowed by tradition and reaffirmed by Darwin's descendants. It is true that as time has passed precursors have been located—Darwin himself was led to attach an account of them as a historical introduction to later editions of the *Origin*—but it has been generally assumed that Darwin arrived independently at his final disclosure, natural selection.

"Of all forms of mental activity," the historian Herbert Butterfield once wrote, "the most difficult to induce is the act of handling the same bundle of data as before, but placing them in a new system of relations with one another by giving them a different framework, all of which virtually means putting on a different kind of thinking cap for the moment." This is precisely what Darwin did when he took the older conception of natural selection and by altering it a hairbreadth created that region of perpetual change, of toothed birds, footless serpents and upright walking apes in which we find ourselves.

The term natural selection has a peculiar history. Under other names it was known earlier within the century, but this is little realized. Darwin in reality took a previously recognized biological device and gave it a new and quite different interpretation. He altered our whole conception of the world in which we live by making use of a principle already known to Lyell and to one other man, a young zoologist by the name of Edward Blyth. Later on in the century the two younger men also became friends. As to whether Edward Blyth ever saw or grasped the connection between his youthful thoughts and the intellectual revolution that came in 1859, we do not know.

At the age of twenty-five Blyth had read Lyell and, impressed by his ideas, begun to carry them a little further. In the British *Magazine of Natural History* in 1835 and again in 1837—the very year that Darwin opened his first notebook upon the species question—Blyth discussed what today we would call both natural and sexual selection. In 1837, Darwin was home from the five-year voyage of the *Beagle,* home with turtles and coral, bird beaks and pampas thistles in his head. That he read *The Magazine of Natural History* around this time, we know from recently discovered evidence. In fact, Darwin, in a somewhat cryptic early letter, tells us so: "In such foreign periodicals as I have seen, there are no such papers as White or Waterton, or some few other naturalists in Loudon's and Charlesworth's Journal would have written; and a great loss it has always appeared to me." Loudon's and Charlesworth's Journal is *The Magazine of Natural History.*

The brief period between the time when Darwin was rowed ashore from the *Beagle* and the opening of his first notebook on the "species question" in 1837, nine months later, contains the real secret of the Darwinian story. We know, for the documents lie ready to our hand, that as late as 1836 and for a short time thereafter, Charles Darwin was still struggling with the problem of the rapidity of organic change. "If one species does change into another it must be *per saltum,*" he writes, "or species may perish."[2] No one has indicated what led the young Darwin to think in this way—a way quite foreign to the views he was later to express in the *Origin.*

With a slight twinkle, Darwin recounted how, after the delivery of his and Wallace's joint papers before the Linnean Society in 1858, Professor Haughton of Dublin had commented "that all that was new in them was false, and what was true was old."[3] Though we need not join Professor Haughton in condemning what was new in the papers, it is still interesting that the learned gentleman seemed to catch echoes of something out of the past. Perhaps, after all, his ear had not played him false.

The fact is that an important shift in Darwin's thinking remains undocumented. It has not even been discussed. In his master work, slow and imperceptible transformations extending over vast ages of time have replaced his early and immature speculations on organic change *"per saltum."* On the origin of the shift, the published notes fall silent. They ask only whether a Henrietta St. Bath would probably answer a letter and give information about a tailless breed of cats near Malmesbury Head.

If it were not for a small series of accidental circumstances, the events we are now about to recount would never have been transmitted to posterity. As it is, one is forced to call upon all the skills of the literary detective in order to make one's way, albeit dimly, into nine months in the lives of two young men nearly a century and a half ago. One of these accidental circumstances was the discovery of Darwin's trial essays on the road to the *Origin of Species.* Written in 1842 and 1844, they had never been printed. They were found long after Darwin's death in a cupboard in the old house at Down. His son Francis published them as the *Foundations of the Origin of Species* in 1909. Since they date into the earlier years of Darwin's work on evolution they yield clues to the past which have vanished in the *Origin.*

The second accident was the rediscovery of Edward Blyth. These seemingly unrelated incidents have a deep bearing upon the problems we have been discussing but, like a puzzle, they have to be fitted together. In that puzzle even the tailless cats will have their place.

II

Twice in a hundred years Edward Blyth has been faintly and somewhat timidly mentioned as a Darwinian precursor: once after his death, though while Darwin was still living, by H. D. Geldart in the annals of an obscure local natural history society, and again in a letter from H.M. Vickers, to *Nature* in 1911.[4] The second discoverer was apparently unaware of the first. In neither instance does the announcement seem to have aroused attention. By the time this resurrection occurred the excitement over Darwin's forerunners had abated. Moreover, in spite of Blyth's importance to our study he was not at this time an evolutionist. To grasp fully his part in early Darwinian events one must know his relation to Darwin.

It is just this point which Blyth's two discoverers failed to pursue, although Vickers, it is true, conversed with Francis Darwin about the matter. This was indeed a strange oversight because Darwin's later volumes are spotted with numerous references to Blyth's work. Anything, therefore, which Blyth might have to say upon species is deserving of the most careful scrutiny.

The paper which excited the attention of Geldart and Vickers appeared in *The Magazine of Natural History* in 1835 a short while before Darwin returned home.[5] It was followed in 1837 by a lengthy second paper which was apparently ignored by the two discoverers of the first document.[6] The two papers are equally important to our study, and I shall make extended use of them. Blyth himself, however, deserves a preliminary word. A year younger than Darwin, he was born, like Alfred Wallace, into straitened economic circumstances. He received only a trade school education but was an omnivorous reader and observer in natural history. He used a small inheritance to purchase a drug business in the town of Tooting in Surrey, now absorbed into greater London. Blyth, it appears, could not abate his passionate interest in zoology sufficiently to achieve business success. Much of his time was spent reading in the British Museum or making the rounds of kindred metropolitan institutions seeking a post more congenial to his interests. He was a born scholar, but he was trapped in miserable economic circumstances and haunted by ill health.

By 1841 he had been advised to quit England for a warmer climate. Having published widely, he was offered a small post as curator of the Museum of the Royal Asiatic Society of Bengal. The next twenty-two years he spent in India, making innumerable specialized contributions to the natural history of southeastern Asia. His health failing, he returned in 1862 to England, where he died in 1873. These are the spare, sad outlines of the life of a potentially great scientist frustrated by the time and the circumstances under which he was born. It is not unlikely that he was reading late in London on that day in 1836 when Charles Darwin, just ashore, was posting fast through the echoing countryside toward Shrewsbury. Edward Blyth was one to remember the color and shape of a darting bird or a fox going over a hedge. He saw things hiding, shifting, changing. He had what today we would call a photographic memory. This memory would contribute, in the end, to the mystery that lies forever between Charles Darwin and Edward Blyth.

III

The idea of selection is, superficially at least, a simple concept. Once the scheme is grasped, multitudinous examples lie ready at hand for illustrative purposes. There are no intricate propositions, no complicated mathematics which can be traced from one individual to another. In simpler forms the idea of selection was known to breeders, and here and there some dim relationship of selection to the struggle for existence was glimpsed by philosophers in both the eighteenth and the early nineteenth centuries. Perhaps it is this fact which has caused scholars to despair of tracing all the peregrinations of the idea through successive hands. By the late eighteenth century it had been touched upon by several writers, including Lamarck, who had, however, skirted the subject without glimpsing its full significance. Its creative aspect beyond the bounds of species had not been grasped—probably both because of religious prejudice and the fact that the length of geological time, along with its successive faunas, remained largely unappreciated.[7]

Because of the enormous prestige of Darwin, as well as this very elusive quality of a simple idea, the discovery that Blyth had written upon natural selection in 1835 fell upon deaf ears. No one, it appears, thought of actually examining Darwin's volumes with Blyth in mind. The present writer sought to do so for just one reason: I failed to comprehend how Charles Darwin could have been unacquainted with the periodical in which Blyth published. It was one of the leading zoological journals of the day. Darwin's friends, Henslow, Jenyns, Lyell, had all appeared in its pages. To assume that a man of Darwin's "prodigious memory and power of abstraction" was unacquainted with writings on the species problem in so prominent an organ as *The Magazine of Natural History* in the period from 1835 to 1837 seemed illogical in the extreme. I set out, therefore, upon a careful and detailed examination of Darwinian materials which might bear upon the problem.

It was obvious that as a first step in my task *The Descent of Man* (1871) and *Variation in Animals and Plants Under Domestication* (1868) were apt to offer more clues than the

Origin itself. Unlike the *Origin* they are extensively footnoted. Furthermore, since the materials for these books dated very largely into the pre-*Origin* days, it was likely that they might contain useful evidence. After all, both volumes were in reality part of the big treatise which Wallace had forestalled by communicating his discovery of natural selection to Darwin in 1858.

My examination of the two volumes led rapidly to some highly interesting discoveries. It became evident that whatever the precise time might be in which he had come upon the series, Darwin had held in his hands, and made use of for scholarly purposes, the volumes of *The Magazine of Natural History* containing Edward Blyth's papers upon natural selection. I list all the references to the magazine in my notes, and it will be observed that they are numerous.[8] They reveal, not alone that Darwin at some point in his life had handled the volumes in question, but also that he had consistently made use of the entire series. Some of the references are to small items and thus show that the volumes had been scanned with great care. In spite of this, and in spite of the fact that Darwin was fond of quoting Edward Blyth frequently upon a considerable range of topics, there is not a solitary direct reference to the two papers which particularly concern us—those of 1835 and 1837.

Charles Darwin had a high opinion of Blyth.[9] In the *Origin of Species* he spoke of him as one "whose opinion, from his large and varied stores of knowledge, I should value more than that of almost anyone."[10] Nevertheless, the two extensive and interesting papers in which Blyth treated of subjects directly pertaining to Darwin's greatest intellectual effort remain, as I have said, unnoted. Even more strange, there is one reference to a paper immediately adjacent to Blyth's paper of 1837.[11] This seeming failure to examine Blyth's papers of 1835 and 1837 is particularly odd when one comes to realize that Edward Blyth has four and a quarter inches of reference space in the index of the *Variation of Animals and Plants Under Domestication*, a space considerably greater than is allotted any other single individual. One begins to get the feeling that something more than chance is at work in this situation. *The Magazine of Natural History* has been obviously utilized at great length by Darwin. We have his own testimony, corroborated by his son Francis, that he "read and abstracted . . . whole series of Journals and Transactions."[12] Yet a man whose work he obviously valued, a man whose name Darwin appears to have taken pleasure in promoting before the public, is represented only by his comment upon specific faunal items. Blyth is restricted to the role of taxonomist and field observer.

It may at this point, however, be still contended that Darwin did not know the files of *The Magazine of Natural History* in his earlier years, and that by the time the citations I have given were utilized he was concerned solely with post-*Origin* subject matter. In this case it would still appear strange that Darwin made no reference to Blyth in his historical sketch appended to later editions of the *Origin.* Allowing this objection to stand, for the moment, however, I went back to the less documented, early phase of Darwin's career in the hope of ascertaining whether any of the surviving papers of that era could throw light upon this curious and absorbing problem. Fragments from the early notebooks of 1836 and 1837, as well as the trial essays, would offer such published clues as might be available. And it was already clear that although Darwin was cognizant of evolutionary ideas before going to sea with the *Beagle,* he did not arrive at a satisfactory solution for organic change until after his return from the five-year voyage.

IV

The notebooks Darwin kept on the voyage contain all manner of stray jottings, ranging from the rise of continental areas to the price of melons. On July 19, 1835, the *Beagle* arrived at Callao in Peru. Darwin's notebook for that day includes a cryptic item: "Smelling properties discussed of Carrion Crows, Hawks, Magazine of Natural History."[13] There can be no doubt that this *Magazine of Natural History* is the one which later was to contain Blyth's first article treating of natural selection. It is evident that more than one issue had reached Darwin by mail at Callao. By the good fortune of his bird references we can identify at least two numbers as those of January and May, 1833

(Volume 6), since these issues contain respectively, an article by Charles Waterton on the habits of the carrion crow and another "On the Faculty of Scent in the Vulture." Darwin's friend Jenyns also had a paper on the principles of taxonomy in this volume. It is now unmistakably clear that Darwin as a young naturalist was as well acquainted with this journal as his colleagues and read it just as assiduously.

In the January 1835 issue Blyth's paper on varieties made its appearance. One year later Darwin was still in Australia and thinking wistfully of home. His final notebook has, unfortunately, not been fully published. Nora Barlow tantalizingly remarks that "he [Darwin] quotes from recent reading and begins a draft of a geological paper, presumably in the leisure of the slow days sailing home across the Atlantic."[14] One would like to know whether Blyth's paper had reached him, but now the last notebooks grow jumbled in time, because Darwin out of long habit continued to carry and use them for two years after coming ashore.[15] Thus, although we cannot identify the precise month in which Darwin first saw Blyth's papers (unless additional unpublished material should throw light on the question), there can be no doubt that he did see them. It is significant, I genuinely believe, that Darwin opened his first notebook on the "species question" in 1837. In January of that year Edward Blyth ventured the beginning of a second paper in which there is comment upon the principle of natural selection. In fact, from 1835 to 1837 there is sporadic discussion upon subjects of an evolutionary cast in *The Magazine of Natural History.*[16] We now come, however, to the crux of the discussion: is it possible to trace in the *Origin of Species* or in the trial essays that preceded it any direct evidence of the influence of Edward Blyth? The answer will affirm the truth of the connection between Blyth and Darwin, but the clues upon which this assertion is based have to be mustered with care. If it had not been for the publication of the *Foundations of the Origin of Species* in 1909 it is unlikely that the dim outlines of the carefully hidden trail would ever have been perceived. This trail begins to be discernible in the Darwin notebook of 1836 with the curious word "inosculate."[17] It is a word which never has had a wide circulation, and which is not to be found in Darwin's vocabulary before this time.

Twice in a single paragraph this word, which means to adjoin, or pass into, is used in connection with evolutionary jottings and speculations. Moreover, these speculations, as will become evident a little later, have a direct bearing upon problems presented in the various Blyth papers. Here I am concerned only to point out that a rare and odd word not hitherto current in Darwin's vocabulary suddenly appears coincidentally with its use in the papers of Edward Blyth.[18] In the *Origin* it survives to make one fleeting appearance as "osculant."[19] Taken in conjunction with other evidence, the rare and mildly archaic character of this word suggests that Darwin acquired it from his reading of Blyth.

It is now necessary to become familiar with Blyth's ideas as expressed in the three papers running from 1835 to 1837 in *The Magazine of Natural History.* At the beginning of this account of Blyth I remarked that he was not an evolutionist. Keeping this in mind one can still see an amazing resemblance between his thought and Darwin's, once one makes allowance for the fact that to Blyth the struggle for existence and natural selection were conservative rather than creative forces. In this restriction Blyth is reflecting the eighteenth-century limitation upon organic divergence which blinded so many early nineteenth-century thinkers.[20] It was Darwin's contribution, of course, that he altered the struggle for existence and made of it a creative mechanism. In doing so, however, he passed by way of the stepping stone of Edward Blyth.

Blyth's paper of 1835, composed when he was just twenty-five years old, betrays immediately the taxonomical interests which were to become so evident in his later writings. It is obvious that he is intrigued by organic change but has not succeeded in breaking out of the current thinking of his period except in three very important particulars: (1) Blyth refused to be engulfed by the mysticism in the quinary taxonomical system of MacLeay and Swainson which was so widely popular in the 1830s.[21] Instead, he sees at the root of any logical system of classification a law of "irregular and indefinite *radiation.*" He observes that "the modifications of each successive type" are "always in direct relation to particular localities, or to peculiar modes of procuring sustenance." [22] (2) Blyth described natural and sexual selection in no uncertain terms and recognized their creative role in the emergence of varieties. He was clearly aware of variation and its hereditary character.[23] (3) In spite of denying that indefinite divergence could be produced in this way, he throws out in one amazing and contradictory passage the suggestion that just as man is able to affect the physical constitution and adaptations of domestic animals, so wild nature might achieve the same success.[24]

It can thus be observed that though Blyth, like Buffon, denied himself, he produced within a short, usable compass an abstract of Darwinian evolution that is remarkably complete. In Blyth's somewhat incoherent and rushing eagerness he threw off, and dashed by, the answers to his own objections. It is an apt illustration of the difficulties involved in escaping from the views of one's own age, particularly when those views are deeply imbedded in one's religious consciousness. In the light of Blyth's papers one can see that a large proportion of Darwin's early effort is devoted to finding a way through the species barrier as erected by Lyell and Edward Blyth, though the latter is never directly mentioned in his text. To realize the extent of this relationship it is necessary, first of all, to get a clear picture of those aspects of Blyth's thinking which are germane to the evolutionary problem which confronted Darwin when he arrived home in 1836. It will be observed that they have a remarkably Darwinian sound.

In the first place, the leading tenets of Darwin's work—the struggle for existence,[25] variation, natural selection and sexual selection—are all fully expressed in Blyth's paper of 1835.

> It is a general law of nature [Blyth observed], for all creatures to propagate the like of themselves: and this extends even to the most trivial minutiae, to the slightest peculiarities; and thus, among ourselves, we see a family likeness transmitted from generation to generation. When two animals are matched together, each remarkable for a certain peculiarity, no matter how trivial, there is also

> a decided tendency in nature for that peculiarity to *increase;* and if the produce of these animals be set apart, and only those in which the same peculiarity is most apparent, be selected to breed from, the next generation will possess it in a still *more* remarkable degree; and so on, till at length the variety I designate a *breed* is formed, which may be very unlike the original type.[26]

Here, of course, we are dealing with artificial selection—with cattle, pigeons, and other domestic forms. After commenting upon sexual selection in herd animals "so that all the young . . . must have had their origin from one which possessed the maximum of power and physical strength; and which, consequently, in the struggle for existence was the best able to maintain his ground," Blyth proceeds beyond purely artificial selection. He applies the principle of selection to wild nature just as Darwin was later to do. The best organized, the most agile will survive and leave the most progeny. He sees selection in nature, however, as a conservative principle "intended by Providence to keep up the typical qualities of a species." The sickly, the ill-adapted must, Blyth contends, soon disappear. The slightest deviation in the coat color of a cryptically adapted form will cause its discovery and destruction.[27]

On this point Blyth stands in about the same position as certain other eighteenth- and early nineteenth-century writers whom I have had occasion to analyze elsewhere, except that he is more conscious of ecological selection.[28] He is cognizant that variant traits can be accumulated by selective breeding under domestic conditions. He is also aware of selection under the struggle for existence in wild nature. But Blyth was primarily a student of living forms. He was a careful observer of the delicate adjustment of life to its surrounding environment. In fact he exaggerated the permanence of these exquisite adaptations which he felt disqualified the animals for any other mode of existence.[29] When circumstances changed, he contended, a species must perish with its locality. Like Lyell somewhat earlier, Blyth had glimpsed the negative aspects of the struggle for existence and the way in which species were eliminated. He failed to see, however, that natural selection was also a potentially liberalizing factor. This is particularly curious because, unlike his contemporary, Swainson, he remained undeceived by the highly artificial numerical taxonomy which was in popular favor at the time. He recognized that species were always "modified in direct relation to particular localities, or to peculiar modes of procuring sustenance."[30] "Just as the surface varies," Blyth observed, "so do its productions and inhabitants." He made out clearly "the reiterate divergence and ramification" of organic relationships.

Blyth's youthful failure, it can now be realized, lay in his provincialism. Perhaps his notion of species would have remained less fixed if he had had Darwin's experience of the new lands. As it was, Edward Blyth tended to see everything except man's domestic productions assigned to its place and kept there by the inexorable force of natural selection. Species boundaries were carefully demarcated. Blyth saw about him the hedge-constricted, precision-cultivated English landscape. Darwin had brought back with him the memories of foreign weeds invading the new world, of introduced animals overrunning the indigenous products of oceanic islands.

For Darwin, fresh from the wild lands, the boundaries of life seemed less rigid and sharp. Nevertheless, he was to spend a great amount of space in the trial essays and the *Origin* answering the arguments of Edward Blyth. He had to seek a way through the mind block of a particular world view; namely, Providence and special creation. Yet, ironically, it was still Blyth who in 1837 suggested the road which passes beyond the purely negative selection to be found in the *Principles* of Sir Charles Lyell. Even in 1835 Blyth had remarked somewhat cryptically that the study of simple variations "properly followed up . . . might lead to some highly interesting and important results."[31] By 1837 he had grown aware that his "localizing principle" was not absolute, that "when a species increases numerically in any habitat beyond what the latter is adequate to sustain . . . either their ranks are mysteriously thinned by what is termed *epizooty,* or an erratic impulse . . . instinctively prompts a portion of them to seek fresh quarters."[32] Blyth noted the dangers which beset those wanderers who pass out of their locality, and he noted the fact that they mostly perish without being able to establish themselves in other places. He was, however, beginning uneasily to sense that if his localizing principle, which is really natural selection "breeding in and in" and transmitting selected individual peculiarities in a single stable environment, were to be broken through by removal of forms to other habitats, unlimited organic change might be possible in wild nature. The situation would then be more comparable to the release offered by man in the controlled breeding of domestic forms.

> A variety of important considerations here crowd upon the mind [confesses Blyth], foremost of which is the enquiry, that, as man, by removing species from their appropriate haunts, superinduces changes on their physical constitution and adaptations, to what extent may not the same take place in wild nature, so that, in a few generations distinctive characters may be acquired, such as are recognised as indicative of specific diversity. . . . *May not then, a large proportion of what are considered species have descended from a common parentage?*[33]

After toying with this intriguing possibility, Blyth felt forced to reject it because "were this self-adapting system to prevail . . . we should seek in vain for those constant and invariable distinctions which are found to obtain."[34] Living species, he contended, would under such circumstances blend into each other—something which does not occur in nature. In the plant world, sea water prevents the germination of drifting seeds carried far from their own locality; animals have homing instincts. "Such antagonist principles obtain throughout creation, whether or not human observation may have yet detected their existence."[35] Blyth's world, in other words, is still that of the English hedgerows, of a stability which gives way only to extinction, never to really marked organic alteration. This stability he came very close to fastening upon the young Darwin. Many pages of Darwin's preliminary essays and of the *Origin* itself are spent in dealing with Blyth's defense of the species barrier. Where Lamarck had seen most so-called extinct species evolving by alteration into living ones, Blyth saw all past forms as extinct and without representation in the present. The compromise solution would have to be sought by Darwin.

As one leafs through Blyth's small papers, however, one is amazed by the ideas which reappear in the trial essays of 1842 and 1844 and which Darwin never altered throughout his life. Besides taking note of natural selection, sexual selection, and the role of hereditary variation, Blyth expressed the view that macromutations would "very rarely, if ever, be perpetuated in a state of nature."[36] Darwin was apparently convinced by Blyth's argument and clung to it thereafter.[37] Blyth believed in the influence of food in the stimulation of variations in domestic stock and, again, we find environmental ideas of this kind lingering in Darwin.[38] Blyth's emphasis on sexual selection in man[39] anticipates the weight given to it by Darwin in *The Descent of Man.*

These are general comments. They suggest the germs, expressed briefly, of what grew to be chapters, even books, in the hands of Darwin. Since our thesis is somewhat startling—namely, that Darwin made unacknowledged use of Blyth's work—the critical historian may still want to ask what interior evidence, what detailed similarities can be used to establish the fact that Blyth is more than a Darwinian precursor, that he is, instead, a direct intellectual forebear in a phylogenetic line of descent. Edward Blyth, in the writer's estimation, belongs in the royal line. He is not an isolated accident. Instead, he is one of the forgotten parents of a great classic. But Darwin's shadow, grown to almost superhuman proportions, lies massive and dark across the early portion of the century. How can one find, even in this similarity of ideas, more than the accidental repetition of like thoughts by different men?

As I have intimated, what exists in the trial essays of Darwin has been almost completely obliterated in the *Origin.* It is the publication of the essays in 1909 that makes it possible to discern the connection I now hope to demonstrate. I have already dwelt upon the obsolete word "inosculate" which suddenly appears in Darwin's notebook upon his return from the voyage and coincident with its appearance in the papers of young Blyth. There is, it emerges, an even more remarkable use of words in a similar context and comparable order which challenges any attempt to explain the situation through accidental duplication. The statistical probability of the same peculiar animals with the same anatomical oddities being listed together in the same approximate order by pure chance is so remote as to be almost nonexistent.

In Blyth's paper of 1835 occurs a statement concerning Ancon sheep. A little later in the same paragraph occurs a list of odd mutations, including "donkey-footed swine, tailless cats, back-feathered, five-toed, and rumpless fowls, together with many sorts of dogs. . . ."[40]

This odd little concentration of mutative types is duplicated in almost the same order in Darwin's essay of 1844. Like Blyth, Darwin is discussing "sports" or hereditary monsters. Like him, Darwin mentions Ancon sheep, rumpless fowls, and tailless cats.[41] It is true that the solid ungular swine and five-toed fowls have disappeared; but they occur in later pages of Darwin's essay. They have merely been dispersed.[42] In the matter of claws, two pages farther on we encounter the phrase, "breeds, characterized by an extra limb or claw as in certain fowls and dogs." In the *Origin* this curious sequence has vanished, though the Ancon sheep is still mentioned.

Blyth, in his discussion of food and its effects on animals, comments that "herbivorous quadrupeds which browse the scanty vegetation on mountains are invariably much smaller than their brethren which crop the luxuriant produce of the plains. . . ."[43] Darwin in turn holds that "external conditions will doubtless influence and modify the results of the most careful selection; it has been found impossible to prevent certain breeds of cattle from degenerating on mountain pastures. . . ."[44]

Blyth, in his discussion of hybridity, recognized dominance (i.e., prepotency) and the possibility of the re-emergence of suppressed characteristics in the third generation.[45] In the essay of 1844 Darwin once again expresses similar views.[46] The use and disuse concept is brought into play by Blyth in his discussion of domestic forms where "an animal . . . supplied regularly with . . . abundance of food, without the trouble and exertion of having to seek for it . . . becomes, in consequence, bulky and lazy

. . . while the muscles of . . . locomotion . . . become rigid and comparatively powerless, or are not developed to their full size."[47] Darwin in his second essay devotes a section to this subject and comes once more to similar conclusions, which are re-expressed in the *Origin.*[48]

Blyth devotes considerable attention to protective coloration and the utilitarian advantages gained by such devices in the struggle for existence.[49] In the course of this discussion Blyth, in his paper of 1835, quoted from Mudie's *Feathered Tribes of the British Islands* the metaphor "grouse are brown heather." In the *Origin* Darwin utilizes the same device, picturing "the red grouse the color of heather."[50] Further on in the same section, and this time in his own words, Blyth speaks of the ptarmigan as "snow in winter." In the same section in Darwin "the alpine ptarmigan [is] white in winter."[51] The discussion of protective coloration is more extended in Blyth, but his and Darwin's views of it are the same. Blyth has a vivid description of the relation of the falcon to its prey. He dwells at length upon the bird's great powers of sight. In the corresponding Darwinian passage hawks are mentioned as "guided by eyesight to their prey."[52] Like Blyth, Darwin then dwells upon the pruning effect exercised by these carnivorous birds in keeping the cryptic coloration of small mammals and ground-dwelling birds, such as grouse, uniform and constant through natural selection. The variant animal is unable to conceal itself successfully and is thus more subject to destruction. Although Darwin's treatment of the idea is not as lengthy as Blyth's, the descriptive material cited above is powerfully suggestive of a direct connection, particularly when it is taken in conjunction with the other evidence I have been at some pains to assemble.[53]

Turning to Blyth's paper of 1837, it will be recalled that in discussing his localizing principle he touched upon the homing instinct in animals. During the course of his discussion Blyth asserted this capacity was "not wholly absent from the human constitution."[54] He referred to the Australian aborigines and other savages as being subject to this "intuitive impulse." Although Darwin almost totally avoided reference to man in the *Origin* it is of interest to note that the homing instinct in man receives attention in both of the trial essays. It is mentioned briefly in the first essay and twice in the second.[55] Here the Australian savage reappears.

The instinctive shamming of death is also mentioned by Blyth as characteristic of certain animals.[56] Once more Darwin treats of it briefly but critically in the essay of 1842 and again in the lengthier essay of 1844.[57] When Blyth's papers are subjected to extended and minute analysis there is no doubt that a few additional items pointing toward a connection between the two men might be established.[58] Even making some allowance for accidental use of the same sources, the effect is cumulative and, in the present writer's view, unexplainable by chance. Furthermore, there is ample evidence that Blyth's restrictions on divarication beyond the species level troubled Darwin and that he was forced to spend considerable time and ingenuity in finding his way around this barrier. After an investigation of this effort, it will be possible to see more clearly than heretofore why Darwin approached the subject of variation in wild nature with such timidity.

V

I have already observed that both Darwin and Wallace had been profoundly influenced by the work of Sir Charles Lyell. We may now add to the list of potential evolutionists affected by that great book the *Principles* the name of Edward Blyth. In comparing the beliefs of Lyell and Blyth in the 1830s we will be able to see more clearly how evolutionary thought was slowly inching ahead during a time which, to many writers, had seemed totally without such developments. It is not surprising that Blyth, like Darwin an eager young scholar, had been impressed by the great learning and elegant sentences of Charles Lyell. What will appear is the one particular in which the work of Blyth passed beyond the thought of Lyell on species. In this one act we can see the closing of the gap in thought between Darwin's forerunners and the author of the *Origin* himself. Blyth is the missing key that makes the entire transition so smooth as to be almost imperceptible.

Like others of the late eighteenth and early nineteenth centuries, Lyell recognized the importance of the struggle for existence, actually used the phrase, and realized that this aspect of the natural world accounted for much extinction throughout past time. "The most fertile variety," he observed, "would always in the end, prevail over the most sterile."[59] Lyell recognized that there was considerable variation in living forms and he commented on the "extraordinary fact, that the newly acquired peculiarities are faithfully transmitted to the offspring."[60] Nevertheless, Lyell did not believe such deviation to be endless. He remarks of domestic animals that "attainments foreign to their natural habits and faculties, may, perhaps have been confirmed with a view to their association with man."[61] He contends that the organization of plants and animals was never absolutely constant but that the plasticity of a given species was limited.

> We must suppose [argued Lyell] that, when the Author of Nature creates an animal or plant, all the possible circumstances in which its descendants are destined to live are foreseen, and that an organization is conferred upon it which will enable the species to perpetuate itself and survive under all the varying circumstances to which it must be inevitably exposed.[62]

The range of variation, Lyell thought, would differ somewhat according to whether the animal had been created to occupy a widely varying environment or a narrowly constricted and uniform one. Thus, the great geologist recognized the possibility that the varieties of a species might diverge in distinct environments or under human selection, but he felt that the preponderance of the evidence favored the view that such divergence was limited in scope—a simple adjustment to insure survival. Recognizing the fact that Blyth had read Lyell, let us now see in what manner Blyth approached the same problem.

First of all it may be quickly observed that although Blyth, too, agreed that there was a limit to divergence, he handled the whole subject with greater clarity and precision than Lyell. Instead of confining himself to vague references as to the effects of climate, temperature and similar factors in determining organic change, Blyth wrestles directly with the genetics of the problem:

> There would almost seem, in some species, to be a tendency, in every separate family, to some particular kind of deviation; which is only counteracted by the various crossings which, in a state of nature, must take place, and by the . . . law which causes each race to be chiefly propagated by the most typical and perfect individuals.[63]

Blyth, long before Darwin had expressed himself on the same subject, had clearly recognized the analogy between artificial and natural selection. "The original form of a species is *unquestionably* better adapted to its *natural* habits than any modification of that form," he asserted, for he saw natural selection as pruning out the least deviation which threatened to unfit the animal for its environment.[64] As a neo-zoologist Blyth had a sharp eye for the incessant variation in nature, but he was also intensely aware of how frequently the deviant form is destroyed. He saw the plunge of the hawk on the animal whose cryptic coloring was imperfect, he observed what happened to creatures who strayed beyond their normal environment. He saw in his mind's eye a world held in a tight dynamic balance:

> It would be easy to point out additional hindrances to the more extensive spread of species of fixed habit, by treating on the fraction which are allowed to attain maturity, even in their normal habitat, of the multitude of germs which are annually produced; and in what ratio the causes which prevent the numerical increase of a species in its indigenous locality would act where its adaptations are not in strict accordance will sufficiently appear, on considering the exquisite perfection of those of the races with which it would have to contend.[65]

Contained in this remark is a full recognition of the Malthusian struggle of population against natural resources, but Blyth believed the cards to be stacked against extended change. It was not an unreasonable position. Species, Blyth had observed, faced extreme difficulties when they wandered out of their normal range.

> Were this self-adapting system [i.e., evolution] to prevail to any extent, we should in vain seek for those constant and invariable distinctions which are found to obtain. Instead of a species becoming gradually less numerous where its haunts grade imperceptibly away, we should discover a corresponding gradation in its adaptations. . . .[66]

Now as a modern student of evolution has recently observed,

> by a paradox the process of natural selection has, over short periods, a conserving effect rather than a modifying one. In a constant environment the great majority of the individuals of a species are very precisely fitted to their habitat, and almost any change from the typical will be a disadvantage: consequently many variant forms are eliminated before they can reproduce. . . .[67]

It was this fact which Blyth had sensed so perspicaciously, but at the same time he had failed to see what lay beyond. His failure was partly the result of inexperience, partly the product of the theological atmosphere of the time. This is characteristically revealed in one of his remarks of 1835 about natural selection. "The same law," he writes, "which was intended by Providence to keep up the typical qualities of a species, can be easily converted by man into a means of raising different varieties."[68]

It should now be apparent why the young evolutionist, Charles Darwin, after discovering Blyth upon his return home from the voyage of the *Beagle,* began momentarily to contemplate the possibility that if "distinct species inosculate[69] so we must believe ancient ones [did] . . . not *gradual* change or degeneration from circumstances, if one species does change into another it must be *per saltum*—or species may perish."[70] Yet with his South American experience before him we see, on the next page, that he is trying to grope around the difficulty raised by Blyth. "Dogs, Cats, Horses, Cattle, Goats, Asses," he persists, "have all run wild and bred, no doubt with perfect success. Showing how creation does not bear upon solely adaptation of animals."[71] It is evident here that he is considering the possibility that animals can respond to change, that Blyth's "localizing principle" is not absolute. "Nothing quite makes sense," writes Darwin's granddaughter, Nora Barlow, of this point in Darwin's career.

One can observe that the reason for this feeling that Darwin is groping lies in the fact that he is trying to find a way around the obstacle raised in Blyth's first paper—the fact that there are distinct breaks between the modern species rather than indefinite passages between them. Lyell, in addition, had contributed to this difficulty paleontologically by his contention that,

> where a capacity is given to individuals to adapt themselves to new circumstances, it does not generally require a very long period for its development; if, indeed, such were the case, it is not easy to see how the modification would answer the ends proposed, for all the individuals would die before new qualities, habits or instincts were conferred.[72]

Blyth had presented Darwin with a possible mechanism of potentially indefinite organic change but had at the same time labeled it as a conservative mechanism ordained by Providence to maintain the stability of species. In 1837, however, as has been noted, Blyth showed a few signs of doubt, hastily suppressed, concerning this notion. It is conceivable that these doubts did not pass unnoticed by the widely traveled Darwin, who had seen Old World forms, intrusive in the New, extirpating or successfully competing with forms supposedly created for a specific habitat. Moreover, Blyth's point of view was indirectly challenged in 1838 by W. D. Weissenborn, who pointed out that numerous species were flourishing in habitats where they were not originally indigenous.[73] The work of the voyager naturalists was making ever more apparent the facts of animal and plant movement and adaptation. Blyth's hedgerow world was in the process of dissolution.

VI

According to Darwin, his recognition of the principle of natural selection came in October of 1838 when he chanced to read Thomas Malthus on population and perceived that the geometric increase of living forms would create a struggle in nature which would in turn promote the survival of advantageous variations. Actually, however, this statement now appears open to some doubt. In the first place, it was not necessary for Darwin to read Malthus in order to realize the intensity of the struggle for existence. Leaving aside Blyth's contribution, mention of it occurs in the writings of Darwin's grandfather Erasmus, in Paley's *Natural Theology,* and in the *Principles of Geology* by Sir Charles Lyell. Even Lamarck mentions it in the *Philosophie Zoologique.* All of these were works Darwin had read when he was young and impressionable. His own son Francis expressed surprise that he should have had to turn to Malthus for inspiration.[74] Furthermore, Francis pointed out what we know to be true, that in 1837 he had already given vent to the essential aspects of the principle.[75]

There are, it is true, a few references to Malthus in the trial essays before the *Origin,* but not to the exclusion of other workers such as de Candolle. Oddly enough, in confiding his great secret to a few of his intimates, Darwin seems to have placed little emphasis upon Malthus. One reads with surprise a letter from Hooker to Darwin written in January 1863. "Did you ever read that painful book, Malthus on Population?" Hooker writes. "I did the other day and was painfully impressed by it."[76] If one turns back to Darwin's letters of the 1840s, one gets the same impression of neglect. Though Darwin wrote often to Hooker, Gray and Jenyns about his work, and about the struggle for existence, Malthus remains unmentioned.[77] In fact, the indices to the first three of the five volumes of Darwin's published letters record no reference aside from the single remark in Darwin's autobiography; and the following two volumes contain only three brief references, two of which concern letters to Alfred Russel Wallace.

My colleague and former student, Dr. Gerald Henderson of

Brooklyn College, has raised an interesting point in this connection. He maintains that *after* Darwin had received Wallace's sketch of 1858 and preparations were made for the joint papers to be given before the Linnean Society, the passage from the essay of 1844 which was selected by Hooker, Lyell and Darwin to be incorporated into Darwin's announcement was the only one in which Malthus was mentioned. Dr. Henderson points out that the famous letter to Gray of 1857 contains no reference to Malthus. Instead, de Candolle, Lyell and Herbert are extolled as Darwin's authorities. It is Dr. Henderson's considered opinion that the passage referring to Malthus was deliberately chosen for presentation to the Linnean Society "because of its correspondence with the subject matter of Wallace's essay."[78] From the time of Wallace's appearance on the scene, Dr. Henderson contends, the significance of Malthus began to bulk larger in Darwin's public declarations about the origin of his views on species.

As I have pointed out, to have referred to Lyell, for example, as the direct source of one's ideas upon evolutionary struggle in nature would have been to quote a man publicly opposed to evolution in support of that doctrine. Since Malthus was active in a quite different field, and was, in addition, the basic source of much of the thinking on the struggle for existence in early nineteenth-century England, it was convenient to have recourse to him as the "authority."[79] It is also possible that Darwin found it easy to fall in with Wallace's use of Malthus partly because a natural rivalry dictated his desire to show he was no less aware of Malthus than Wallace was. This curious chain of events had, in any event, the effect of obscuring ever more deeply the real origin of Darwin's evolutionary system. Some of Darwin's hesitations, long delays over publishing, and almost neurotic anxiety can now perhaps be better understood. He had his secrets, and, as I hope to show a little later, he had his justification for them.

VII

We have previously observed that the main difference between Blyth and Darwin lies in the fact that one was a special creationist and the other an evolutionist. The thing which strikes one as unusual in this connection is that both men were reasoning from the same principle: natural selection. Blyth, however, had discovered its short-time stabilizing effects. He was, as we have seen, still laboring under the theological aspects of the argument from design. His emphasis is largely upon natural selection as a providential localizing principle confining animals and plants to their proper sphere of activity—the place for which they were created. Only man, through the intelligent use of artificial selection, is in a position to multiply varieties and perpetuate peculiar forms.

Although Blyth, like Lyell, appears to have regarded the changing environments of the past as resulting simply in extermination,[80] he does hint cryptically in a paper of 1838 that the only evidence for the continued existence of the "creative energy" lies in the "results afforded by the study of fossil remains."[81] Darwin, calling upon his extended South American experience, was not impressed by the hedgerow-everything-ordained-to-its-own-place philosophy of Blyth. Taking such hints as were provided by his own experience and the timid insights which Blyth failed to pursue, Darwin denied that everything in the animal world must perish with its locality. For a time he choked, as we have seen, upon Blyth's and Lyell's argument that adjustment would come too slowly to fit an animal to survive in an environment for which it might not, in the beginning, be as well fitted as the indigenous occupants of the ground.[82] For this reason he considered briefly the possibility of macro-mutative leaps as a way through this particular difficulty—adjustment, in other words, of a more instantaneous variety. He soon saw, however, that this was both scientifically dubious and unnecessary. From the time of the first essay he begins to answer the question of how a transitional form can subsist until its adaptations have been perfected. "I would rather trust . . . pure geological evidence than either zoological or botanical evidence," Darwin once remarked to Hooker.[83]

It is here that he can be seen taking boldly to a course at which Blyth only hinted. For Darwin altered Blyth's scheme by openly attaching to it Lyell's infinite time and changing geological conditions. To it he added the view that the species barrier was an illusion created by the imperfections of the geological record and the shortness of human life. Organic change occurred at a rate which was imperceptible. As a matter of fact, men tended to forget even their transitional domestic varieties once a new breed became established. "To be brief," Darwin wrote to Asa Gray in 1856, "I *assume* that species arise like our domestic varieties with much extinction."[84]

Long before this, however, in the first essay of 1842, he can be seen marshaling his arguments for the purpose of breaking through Blyth's conservative "localizing" system. In summary form his points can be stated as follows:

(1) The conquest of an indigenous fauna by introduced organisms "shows that the indigenes were not perfectly adapted,"[85] that is, there is no such thing as an animal providentially ordained to occupy a given environment. Adaptation is, and will always be, only a relatively successful achievement often broken in upon and destroyed. Thus there is no real barrier to continued change.[86]

(2) Oscillations of sea level on islands and continents create shifting conditions—isolation, faunal migrations, etc. This leads to opportunities for *new* selection of *new* characters, not just the *conserving* selection of Blyth.[87]

(3) New conditions, Darwin maintained, increase the tendency for mutations to appear. This happens under wild conditions in a form analogous to what Blyth contended was the case among domestic animals.[88] Similarly migrations of faunas and floras into new environments would promote mutation.

(4) Darwin argued cogently that *no* environment is completely static, and therefore renewed selection is going on even under superficially uniform conditions.[89]

(5) Darwin invokes a law of succession to explain why the living organisms of a continental area ordinarily show the closest affinities to the extinct life of the same region. This anatomical relationship between the past and the present is only explainable on the basis of evolution.[90]

(6) Finally, in the *Origin of Species,* Darwin devotes an entire section to the formidable point raised by Blyth as to the

rarity or absence of transitional varieties between the various modern faunas. It was Blyth's argument, based on modern observation, that transitional varieties could not have subsisted. Referring to his "opponents" with deliberate vagueness in Chapter 6 of the *Origin,* Darwin confesses, just as could be deduced from an examination of Blyth and the notebooks of 1836, that "this difficulty for a long time quite confounded me."[91] Nevertheless, he came to believe the difficulty could be explained. We tend, Darwin argued, to think too much in terms of recent climate and geographical gradation and then expect life to grade itself as imperceptibly. Actually life is historical. It may intrude into new areas. Species compete and sharply limit one another's expansion. Intermediate varieties, therefore, tend frequently to contract their range and disappear. Thus Darwin, after long study and analysis, broke through the seemingly durable species-barrier which Blyth had, paradoxically, created upon the basis of natural selection.

All of these points and more were, by degrees, shaped into the final perfected argument of the *Origin of Species.* Natural selection, as can now be seen, was used first as a conservative teleological device for maintaining the stability of the natural world. This was its primary function in the hands of Edward Blyth. Charles Darwin, the vastly gifted synthesizer, the perceptive coordinator of materials from diverse sciences, remains the massive figure he has always been.

One last question remains to be considered before turning to the forces which in some degree may have motivated Darwin's behavior. It has been observed that he chose to ignore publicly the papers of Edward Blyth. Is it possible that the man who once wrote, "All my notions about *how* species change are derived from long-continued study of the works of . . . agriculturists and horticulturists," could have been secretly aware of any other precursors?[92] After the Blyth episode one cannot help but wonder. On page 78 of the 1844 trial essay on the road to the *Origin* occurs the following observation:

> In the case of forest trees raised in nurseries, which vary more than the same trees do in their aboriginal forests, the cause would

> seem to lie in their not having to struggle against other trees and weeds, which in their natural state doubtless would limit the conditions of their existence.

When one turns to the *Origin* itself one finds that this statement has vanished. In Patrick Matthew's book *On Naval Timber and Arboriculture,* which fully anticipates the idea of natural selection, there is to be found, however, this intriguing remark:

> Man's interference, by preventing this natural process of selection among plants, independent of the wider range of circumstances to which he introduces them, has increased the difference in varieties particularly in the more domesticated kinds.[93]

Darwin's comment, in other words, sounds as though based on Matthew—at least I am unaware of any other convenient source. So strongly did I feel this to be the case that, upon encountering Matthew's statement, I looked into the *Variation of Animals and Plants Under Domestication,* curious as to what one might find there. By 1868 Matthew's book had been publicly called to Darwin's attention. There was thus no reason why he should not refer to *Naval Timber and Arboriculture.* In the second volume of Darwin's work one comes immediately upon the following reference under "Matthew":

> Our common forest-trees are very variable, as may be seen in every extensive nursery-ground; but as they are not valued like fruit-trees, and as they seed late in life, no selection has been applied to them; consequently, as Mr. Patrick Matthew remarks, they have not yielded different races. . . .[94]

This statement appears to be a shortened version of the 1844 comment placed in a new setting, with some additional comment from another page of Matthew added.[95]

The re-emergence of this discussion of the variability of forest trees in nurseries suggests that Darwin *was* aware of Matthew by 1844. The fact that Darwin, as has been seen, speaks forcefully of his long and persistent searching of horticultural and agricultural sources, makes it less easy to accept his ingenuous disclaimer that "one may be excused in not having discovered the fact [i.e., natural selection] in a work on Naval Timber."[96]

Another intriguing point arises in connection with Matthew —the question of where Darwin got the idea for his phrase "natural selection." Such phrases, though easy to appreciate after their appearance, are deceptive in that they often emerge from at least some similar expression, rather than being totally original. Blyth does not use the expression nor, with one exception, does it occur in Darwin's essay of 1842.[97] It does appear in the second essay of 1844. The nearest thing to Darwin's usage occurs in the passage from Matthew which I have already quoted. In it Matthew speaks of *"this natural process of selection."* Darwin uses the expression "natural means of selection"[98] in the essay of 1842, and again in the title of Chapter II in the essay of 1844. Later in the text, however, the term appears as we know it today. The similarity of Matthew's phrase cannot by itself, perhaps, be regarded as conclusive evidence.[99] When it is taken in conjunction with our other material, however, there can be no doubt that Matthew deserves attention.

VIII

We have now come, through a very tedious and detailed effort, to a question of considerable magnitude. Why did Darwin, a figure of such stature in science, feel impelled toward this grudging and secretive behavior? Many pages of his biographies are devoted to his magnanimity, his friendliness, his lack of pretense. On the other hand, it is well known that he had his moments of indifference toward his forerunners.[100] He was capable of saying in his autobiography that he had never encountered a single naturalist who entertained doubts on the permanence of species, although in a letter to Hooker in 1847 he had commented jovially, "I see you have introduced several sentences against us transmutationists."[101]

Attempts to explain some of these paradoxes of character have been legion. There have been also the complications introduced through the unconscious process of myth-making, the desire, in other words, to keep this man and his discovery inviolate—a unique act of genius without precedent and without precursive steps. There has thus arisen a tendency to see Darwin's forerunners as having no relation to his own accomplishment. They are dismissed, as Darwin was inclined to dismiss his own grandfather, as "part of the history of error," as speculative, as lacking in facts. The irony in this situation lies, of course, in the fact that many of the ideas Darwin was later to use came from the researches of these very men. Lamarck, for example, observed the struggle for existence and recognized the significance of vestigial organs before Darwin. Are we to say, because Darwin postdates the earlier workers and had more information available to him, that their observations only became facts in his hands? It has been contended that Darwin sometimes thought in this fashion but at least had the grace to confess that he had little feeling for history.[102] There is no reason over one hundred years later to embrace the same fallacy.

Charles Darwin, like every other worker in the field of science, used the knowledge and the accumulated stores of information of his predecessors. To their efforts he added his own vast resources and the originality of a powerful, far-reaching mind. R. Taton, in his stimulating book, *Reason and Chance in Scientific Discovery* (1957), points out that new discoveries, particularly where some degree of synthesis is involved, are often made by one who has knowledge derived from more than one field of thought. Certainly this is true of Charles Darwin. He took the providential "localizing principle" of a neo-zoologist like Blyth and added to it the infinity of geological change. He drew also, from the analogy of domestic breeding, the fact that change was potentially endless and, given time, could carry organisms far beyond the restricted habitats to which Blyth and his fellow naturalists had confined them.

Darwin's solution, in essence, was merely another way of looking at the same set of data, but it was the dispassionate observation of a man on a height to which no one else had climbed. Nevertheless, Darwin himself once spoke appreciatively of the contributions of the less gifted workers in the field of science. Here again one catches a momentary glimpse of

Darwin's ambivalent psychological behavior—his curt dismissal of those who had come close to his pet theory, and yet again his remorseful praise of the "little man" in science.[103]

Charles Darwin, often depicted as a simple, forthright man, was in reality an enormously complex human being. Consider, for example, the unconscious irony to be found in these remarks to Lyell in 1860: "I have had a letter from poor Blyth of Calcutta, who is much disappointed at hearing Lord Canning will not grant any money. . . . Blyth says (and he is in many respects a very good judge) that his ideas on species are quite revolutionized. . . ."[104] It is within the parenthesis that the irony lies, but it may be suspected that Darwin was only speaking what to him was an experienced truth.

Irrespective, however, of various recent attempts to psychoanalyze Darwin and to consider the psychosomatic aspects of his long illness—his reluctance to meet the public, his disturbing stomach symptoms, his anxieties—there still remain the social imponderables of the time. The French Revolution had had a disastrous effect on English science. Alexander Maxwell in his *Plurality of Worlds* (1817) speaks of the "infidel notions and absurdities which abound in Buffon, Hutton, Playfair and many of the French writers upon the same subject."[105] He castigates "astronomical fable" and "visionary speculation." Similarly, the zoologist William Swainson speaks critically of what is actually the evolutionary approach. "A cold, ill-concealed spirit of materialism, or an open and daring avowal of wild theories, not more impious than they are absurd, attest . . . the infidelity that attaches to some of the greatest names in modern zoology which France, or indeed any other country has produced."[106]

Evolution was suspect as French atheism. English naturalists discussing the species question disavowed Lamarck with the ritual regularity of communists abjuring deviationist tendencies today. From this standpoint it is likely that the eagerness of Darwin to disclaim any assistance from Lamarck was little more than the common English reaction of the time, however ungracious it may now appear. It would seem that the subject of evolution itself offered difficulties to a man engaged in reopening what was regarded by the majority as a closed episode in biology. Darwin *was* confronted by a genuinely unusual problem. The mechanism, natural selection, by which he hoped to prove the reality of evolution, had been written about most intelligently by a nonevolutionist. Geology, the time world which it was necessary to attach to natural selection in order to produce the mechanism of organic change, had been beautifully written upon by a man who had publicly repudiated the evolutionary position.

Here was an intellectual climate in which men were violently opposed to evolution as godless and immoral. Here was the germ of the idea itself, the struggle for existence, regarded as a mere pruning device for keeping species up to par. "The most important views," Darwin reminisced in his old age, "are often neglected unless they are urged and re-urged."[107]

At some time very soon after his return from the voyage of the *Beagle* Darwin apparently made a decision: the evolutionary idea would have to be launched again as something totally new, and it would have to be launched with a massive accumulation of demonstrable facts to bolster it. In the atmosphere of that time, to have footnoted one's ideas as derived either from French sources or from men who had already taken a different stand on the same evidence would have led to little but embarrassment. Darwin settled down to a long period of fruitful labor and watchful waiting. His reputation as a sound, conservative naturalist was growing. He watched the critical attack on Chambers' *Vestiges of the Natural History of Creation*—an attack made more ferocious by religious bigotry.[108] He watched and was silent while Huxley violently condemned Chambers' book. Throughout the whole time his own huge treatise was growing. Darwin had an almost pathological patience. In the end it came close to destroying him. Only the fact that Wallace chanced to send his own discovery to Darwin instead of to a journal prevented his anticipation of the *Origin of Species*.

If one considers this peculiar set of circumstances one can realize, objectively, that these conditions are not those of today and should not, perhaps, be judged by today's standards. During

Darwin's final year, indeed, one observer put the matter quite succinctly. Speaking of Darwin's predecessors, he wrote:

> To have relied in any way on their authority when Mr. Darwin's book was first published might well have increased the mountain of prejudice against his views without in any way relieving the weight of ridicule that lay upon theirs. When the whole scientific world had been stirred to its foundations and when the whole world almost had been roused into paying attention to science . . . then when it could best be done, Mr. Darwin turned ridicule into renown, and made all who could even remotely claim to have anticipated or shared his views participators of his fame.[109]

This comment by a contemporary who apparently sensed some of Darwin's difficulties is interesting and pertinent, but certainly excessive in its praise of Darwin's late atonement to his forerunners. In the first place, Darwin's gesture is meager and confused, as Professor C. D. Darlington pointed out some years ago when he commented, "There are many, like Bateson, who complained at least privately, of Darwin's disregard of historical propriety or historical knowledge."[110] There is every indication that Darwin found the historical introduction which was included in the *Origin* a painful task. It would seem that beneath the political necessities which undoubtedly had early contributed to Darwin's reluctance to review his forerunners, there was a genuine and understandable hunger to possess the theory as totally his own.

From a tolerance of what he used to term "speculative men," from a willingness to call himself "a gambler who loved a wild experiment," Darwin, as he became by degrees a British institution, appears to have sobered into what his granddaughter, Nora Barlow, observes as "his repudiation of those who spin their theories without the constant discipline of factual detail."[111] In this attitude, it has been contended, has lain much of his rejection of his forerunners.

It is true that the *Origin* was originally conceived to be an abstract of a larger work and that Darwin wrote to at least one correspondent advising him not to expect footnotes.[112] One can observe, however, that if there had been any pressing desire on his part to remedy this situation, it could easily have been done in later editions. One can only conclude that Darwin was solitary and elusive beyond even what his family has recorded. "This sense of solitariness," wrote Charles Cox in 1909, "followed him to the end of his life and was, no doubt, an important factor in the formation and preservation of his extraordinary individuality and faith in his own powers."[113]

As one looks back over the curious and intertwined history of Darwin and his associates, one is struck more and more by the smooth, unbroken evolution of the natural selection concept from the time of the eighteenth-century writers onward. An enormous body of myth has obscured this process. Today the number of eulogies, addresses and similar encomiums is burying ever deeper the true story of the past. The flash of genius, the master stroke, arouse our pride in human achievement. Few of us want to learn that many less fortunate men toiled to erect the edifice later to be known as Darwinism. Fewer still will accept the fact that Darwin was a man among men, who yearned hungrily for the approbation of the world—perhaps, as has been intimated by psychologists, as a compensation for the doubts of a domineering father.[114] Yet it is this man, not the bearded idol, who is one of us if we would but see it.

Charles Darwin was an explorer who found no answer for the questions that hounded him throughout the five years of a great voyage.[115] Upon his return to London he combed journals with tenacity until he found what he was seeking. In doing so he broke through the invisible barrier before which other men—even Blyth—had hesitated, the fixity of living forms. Once, in a letter to Hooker, he commented that it was a pity that scientific men read so little. "I have often thought," he insisted, "that science would progress more if there was more reading."[116] This was the man, the plodding man with the enormous unyielding patience and uncanny insight, who piled up folders of evidence until the world capitulated. In Darwin, the boy in every one of us is vindicated. Perhaps that has been part of our secret, if confused, adulation. For Darwin, we thought, was the youth who failed at school and ran away to sea only to return in triumph with the secrets of Sinbad's cavern.

This dream need not die because the sailor runaway was helped by others, or because he picked up doubloons of gold secreted in the treasure chests of unread journals. Rather, it should be a lesson to all of us. Sir James Paget once remarked that Darwin's volumes "exemplified in a most remarkable manner" his power of "utilizing the waste material of other men's laboratories." One might venture the observation, in the light of our present study, that he was equally adept in the utilization of those stray hints which fell from other men's minds, but which in his own head underwent a marvelous transformation.

The finest vindication of Charles Darwin, man and voyager, lies in the fact that Edward Blyth (who also fled England, and, traveler-naturalist that he was, bargained once for eighteen tigers in the river port of Lucknow), respected Charles. "He is a very clever, odd, wild fellow," Darwin once wrote in turn of Edward Blyth, "who will never do what he could do, from not sticking to any one subject."[117]

They knew each other well, those two. Blyth, so far as the published materials inform us, chose never to remember that he had, in the 1830s, written of things which, however differently, had swept the world in 1859. Being both generous and modest, perhaps he never saw any relation between his youthful cogitations and the great change in human thinking which ensued a quarter of a century later. (Blyth was quick to notice, however, the significance of Wallace's paper of 1855 on the law of succession, and mentioned it to Darwin. This suggests he was following evolutionary discussions in this later period with interest.) Or perhaps he thought the old conception common. One cannot help but wonder. We know, certainly, that Blyth was several times a guest at Down and that many unpublished letters passed between him and Darwin.[118]

It is well to end upon and to respect the two men's mutual silence. But let the world not forget that Edward Blyth, a man of poverty and bad fortune, shaped a key that dropped half-used from his hands when he set forth hastily on his own ill-fated voyage. That key, which was picked up and reforged by a far greater and more cunning hand, was no less than natural selection. At that moment, probably in 1837, the *Origin* was born.[119] When Blyth died in 1873 there was found among his papers a fragment of a work which he was preparing "On the Origination of Species."[120] It was not, his literary executors opined, worth publishing. It was derivative. In truth it was the dry seed husk fallen from what had grown to be a great tree.

Darwin, Coleridge, and the Theory of Unconscious Creation

AFTER the *Origin of Species* was published in 1859, Darwin wrote to the Reverend Baden Powell, "If I have taken anything from you, I assure you it has been unconsciously."[1] This was in response to a letter in which Powell had reproved Darwin for not referring to one of his works. A man whose mind was stuffed with the multitudinous lore of both incredibly wide reading and oceanic personal experience, Darwin openly proclaimed himself a millionaire of odd and curious little facts. There floated in his vast memory the tortoises and lizards of islands under tropic suns. He had dug for fossil bones in Patagonia, and climbed Andean peaks in solitude. Navigators and albatrosses were part of his experience. In this he resembled another nineteenth-century genius—the poet Samuel Taylor Coleridge, opium addict, creator of *The Rime of the Ancient Mariner, Kubla Khan,* and the weird moonlit fragment *Christabel.* Now, by the vicissitudes of fortune, the two men have become posthumously involved in a stormy controversy among twentieth-century biologists.

"For over a century," writes the critic Max Schulz, "the tradition has been that they [Coleridge's poems] defy analysis because the best of them are enchanted records of unearthly realms peopled by Mongol warriors, old navigators, albatrosses, and Lamia witchwomen. . . ."[2] Students of the subject have been until recently loath to perceive the conscious craftsman behind the dreamer. The tendency has been to accept "the sacred river," the sunless sea of dream as the primary source of Coleridge's inspiration. Out of these misty depths, according to entranced critics, were drawn in poetic ecstasy fragments of travelers' tales transmuted forever in the subconscious mind of the poet. The public appeal of this romantic interpretation of great poetry is tremendous. It flatters our imagination and our conception of the mysterious life of the literary artist.

In the case of Darwin, the public imagination was, and still is, caught by the symbol of a great voyage, the voyage of the *Beagle.* Thus the legend of the unconscious, the role of the "sacred river," was destined to leave the realm of poetry and enter the field of science. The floating fragments from Darwin's sacred river need no longer be assigned place, name, or priority. They had become the property of genius, they had entered the dark domain of demonic creation. As it is in literature, where historic footnotes are not demanded of the poet, so now it was about to become in science. Charles Darwin was to be elevated beyond giving an account of even partial priority as that rule applied to other men.

Darwin's own excuse of the "unconscious" has been increasingly used in recent years by defenders of the great biologist in considering the problem of Darwin's meager attention to his predecessors. If Darwin "unconsciously" borrowed material without acknowledgment, some scholars imply that no blame can be attributed to him. Rather, they frown upon those historians of science who persist in probing beneath the insights of genius in order to seek the sources of their inspiration. Yet we must still ask, was the one man who should know where he got the idea of his famous theory unconscious of where he got it? Or did he consciously draw a veil over one predecessor in particular, rationalizing, perhaps, as he is known to have done on one occasion, when he implied that the man who successfully convinces the public of a new idea deserves all the credit which may accrue to him.[3]

The theory of the "unconscious" has been emphasized by Darwinian defenders particularly following the publication in 1959, in the *Proceedings of the American Philosophical Society,* of my article exploring the possible role played by Edward Blyth in the formulation of the theory of natural selection.[4] The publication of Darwin's Notebooks on Transmutation of Species,[5] in 1960, showed clearly that Darwin was aware of Blyth's writings on natural selection. In the second notebook,[6] never intended for publication, reference is made to the article of 1837 in which Blyth writes, after having earlier described the conservative effects of natural selection: "May not, then, a large proportion of what are considered species have descended from a common parentage?"[7]

Several who are unwilling to credit Blyth with influencing Darwin refuse to quote this line of 1837—the very year that Darwin conceived of the role of natural selection in evolution. Sir Gavin de Beer, who edited the notebooks, footnotes Blyth's article as mentioned by Darwin in the second notebook, but fails to point out its obvious import. In a brief discussion of Blyth in the introduction to the first notebook he at first remarks that Darwin probably owed nothing to Blyth so far as the construction of his theory is concerned.[8] He confesses, however, in the same paragraph, that "there is nothing improbable in his [Darwin's] having copied some from Blyth." He then cautiously concedes that "Darwin (and others) may have been wrong in thinking that he owed him [Blyth] or them nothing on this score." Four years later, in 1964, in his biography of Darwin, de Beer again retreated from a direct confrontation of the full nature of Blyth's speculations when he says that although Blyth "had been playing with the very tools that Darwin so successfully used, it is difficult to see that Darwin was indebted to him, for his conclusions were the exact negation of what Darwin was trying to prove."[9] Concerning this statement it is of interest to note that George Wald, writing on "Innovation in Biology," in the *Scientific American,* remarks that "all great ideas come in pairs, the one the negation of the other, and both containing elements of truth."[10] Edward Blyth, as we have seen, in a moment of insight glimpsed momentarily both faces of natural selection. It was enough to give an astute mind like Darwin's the clue that he was seeking.

Theodosius Dobzhansky, who uses the "unconscious" theory to explain Darwin's failure to acknowledge his predecessors, accepts the fact that "the fundamental premises of the theory of natural selection are contained in Blyth's essays," but maintains that Darwin might have been mistaken about the sources of some of his ideas, and his thinking process might not have been wholly free of subconscious components.[11] Dobzhansky's theory of the "subconscious components" as the probable cause of Darwin's omission of credit to Edward Blyth should be considered carefully because it gives us an opportunity to explore one interpretation of the creative processes of genius.[12]

Is the origin of every idea which crosses our minds always clear to us?" Dobzhansky asks. Nobody, he contends, has a perfect memory and is always aware of his thinking processes.

> Probably everyone is familiar with the feeling that an idea which arises in one's mind, or a phrase which emerges from one's pen, have been met with somewhere, but one cannot recall just where or when. This feeling is sometimes justified but perhaps more often illusory. Might not even Darwin have been mistaken about the sources of some of his ideas?[13]

It is the mystery of the creative process in the mind of genius which the discovery of Edward Blyth forces us to face, argues Dobzhansky. In Darwin's defense, Dobzhansky cites John Livingston Lowes' well-known study of Coleridge, *The Road to Xanadu.* Dobzhansky feels, as did Lowes, that "it is not illegitimate to compare the creative processes of a poet, Coleridge, with those of a scientist, Darwin."[14] Since the creative work of a poet and that of a scientist are not fundamentally different, Dobzhansky maintains that Darwin had little more awareness of the soil in which his theories grew than did Coleridge of the sources of his poetry.[15] Unfortunately for Dobzhansky's reliance upon this interpretation of Coleridge, however, later critics no longer see the great romantic poet as purely an inspired somnambulist. In the first volume of Kathleen Coburn's edition of Coleridge's notebooks, says Schulz, "the notations of ideas and images for future poems recorded in them reveal a mind knowing where it is going and moving purposefully toward that goal."[16]

Could Darwin have been unaware of the fact that he had read and utilized the articles written by Edward Blyth on the subject which he was later to claim completely as his own discovery? Lowes has written:

> The "deep well of unconscious cerebration" underlies your consciousness and mine, but in the case of genius its waters are possessed of a peculiar potency. Images and impressions converge and blend even in the sleepy drench of our forgetful pools. But the inscrutable energy of genius which we call creative owes its secret virtue at least in part to the enhanced and almost incredible facility with which in the wonder-working depths of the unconscious the fragments which sink incessantly below the surface fuse and assimilate and coalesce. The depths are peopled to start out with (and this is fundamental) by conscious intellectual activity, keyed, it may be, as in Coleridge's intense and exigent reading, to the highest pitch. Moreover (and this crucially important consideration will occupy us in due time), it is again conscious energy, now of another and loftier type, which later drags the deeps for their submerged treasure, and moulds the bewildering chaos into unity. But interposed between consciousness and consciousness is the well. And therein resides the peculiar significance of such a phantasmagoria as lies before us in the [Coleridge's] Note Book, the seemingly meaningless jumble of which we have tried to grasp.[17]

In striking contrast to this view, however, Werner W. Beyer, another Coleridge authority, emphasizes the role of the conscious and deliberate in the creative process. Writing in *The Enchanted Forest,* which the author says begins where *The Road to Xanadu* ended, Beyer states that Lowes' stress upon the importance of the unconscious "has given such wide currency to the concept of unconscious metamorphosis that its conscious counterpart has threatened to be ignored."[18] Beyer presents interesting evidence of Coleridge's use in his poetry of a crucial source until now undetected. This deals with the part that

C. M. Wieland's tale *Oberon,* translated by Coleridge about November 20, 1797, seems to have played in the genesis of *The Wanderings of Cain, The Rime of the Ancient Mariner, Christabel,* and *Kubla Khan.* Beyer has no doubt that Coleridge was aware of Wieland's poetry as a source of his own, since Coleridge translated the tale, and wrote this fact in a letter to Joseph Cottle.[19] Lowes, himself, in a letter of November 24, 1939, wrote to Beyer concerning this newly produced evidence: "Why, in view of the fact that on p. 243 of *The Road to Xanadu* I referred to S. T. C.'s flat statement that he was translating *Oberon,* I didn't go farther, I can't, to save my soul, imagine!"[20]

It was *conscious* judgment, Beyer insists, that led Coleridge to his discoveries of the potentialities of *Oberon.* More than Lowes suspected,

> . . . conscious and unconscious appear to have collaborated and interpenetrated in the genesis of the fabulous ballad [*Rime*]. *As others have thought, the deliberate, purposive, and volitional appear to have played a far greater role in the complex process of discovering and envisioning, assimulating and transforming the multifarious stuff for its fabric and form. . . .* Oberon *makes it clear, I think, that the genesis of the* Ancient Mariner, *which it generously abetted, was not so largely a product of the subconscious as Lowes assumed.* [21] (Italics L.E.).

Beyer also cites another Coleridge authority, R. C. Bald, who,

> . . . after a study of the later notebooks [of Coleridge], similarly stressed what Lowes had seemed to minimize; the *conscious* element in the creative process, the deliberateness of Coleridge's reading, for purposes of poetry, and the recency of some of it which therefore could not have been long submerged in the subconscious. . . .[22]

Since it has been noted that the creative processes of a poet (Coleridge) may be compared to the creative processes of a scientist (Darwin), it is impossible not to see the parallel between these two geniuses in their working processes. *Oberon,* Beyer remarks, appears to have provided a scenario that Coleridge found

> . . . adaptable, kindling, and of high "symbolic potential." . . . It seems to have provided many materials, too, but, more important, to have served as a flexible form or matrix to help organize the richly diverse ingredients drawn from innumerable other sources and experience previously unrelated. And the conscious guidance it evidently afforded appears to put a somewhat different complexion on the story of the genesis of the great ballad, and at the same time to shed new light on various obscure passages.[23]

Coleridge once made the remark that men are caterpillars, very few of whom succeed in successfully transmuting themselves into butterflies. To introduce a bit of modern knowledge, one might observe that the caterpillar possesses glands in its head which at the proper moment assist it to make that beautiful transformation. In the case of men, even brilliant men, some outside incident, some catalytic agent concealed in the environment, may be the initiator of the transformation we call "creativity" or "genius." The man must be receptive, his mind afloat perhaps with the random forms which contain an unevolved future. It is then that the hidden key to the locked secret must be found. Otherwise the potential inspiration may drift past unrecognized into oblivion.

As *Oberon* was such a stimulus, a scenario for Coleridge's great poems, Blyth's articles on natural selection were conceivably Darwin's scenario, containing as they did the full series of stepping stones over which, as I have elsewhere pointed out, Darwin passed on his way into the new world of organic novelty. Lowes remarked as much of Coleridge: "[In 1797] . . . a vast concourse of images was hovering in the background of Coleridge's brain, waiting for the formative conception which should strike through their confusion, and marshal them into clarity and order."[24]

In October or November, says Beyer, "the young poet-in-waiting discovered *Oberon,* and speedily began translating its teeming kaleidoscopic scenes—scenes which were *as if made for that service.*"[25] (Beyer's italics.) Coleridge may be compared to the young scientist Darwin, home from his voyage, freshly impressed with new lands and unknown creatures. In the case of Coleridge, there exists a written admission that he was translating Wieland's *Oberon.* In the case of Charles Darwin, although he at no time mentioned Edward Blyth's ideas on natural selection,[26] interior evidence such as I produced in 1959, and which also appears in Darwin's Second Notebook on Transmutation of Species, shows that he was fully aware of the papers which contained these ideas.

Dr. Gerald Henderson of Brooklyn College has kindly allowed me to utilize additional evidence from his own recent unpublished investigations at the Cambridge University Library. Darwin's personal volume of *The Magazine of Natural History* of 1837 reveals annotations on Blyth's paper in Darwin's own hand. Moreover, a set of Darwin's page reminders which include Blyth's paper has been pinned to the inside of the back page. I will not encroach further upon Dr. Henderson's researches except to reiterate that Darwin knew and studied the 1837 paper he was never to mention in print. The volume at Cambridge is a presentation copy given to Darwin by Edward Charlesworth, its then editor.

Coleridge himself said: "Though [my mind] perceives the *difference* of things, yet [it] is eternally pursuing the likenesses, or, rather, that which is common [between them]."[27] The following lines were written concerning Coleridge; they could also have been written about Charles Darwin: "We have to do, in a word, with one of the most extraordinary memories of which there is record, stored with the spoils of an omnivorous reading, and endowed into the bargain with an almost uncanny power of association."[28]

J. B. Beer adds that ". . . from sources so widely separated in space and time, Coleridge had often elicited an image or a phrase which was infinitely richer than the sum of its source."[29]

Coleridge, far from defending the bathos of unconscious discovery, remarked with surprising practicality even in the midst of a commentary on sleep: "O then what visions have I had, what dreams—the Bark, the Sea . . . Stuff of Sleep &

Dreams, *& yet my Reason at the Rudder.*"[30] (Italics L. E.) He lists sensations, items of interest in his notebooks, just as does Darwin, but this is the stuff of poetry, not poetry itself, just as Darwin's associations of ideas and lists of sources are the stuff of science but not the completed act of reason.

As Professor Schultz has indicated, Coleridge was not engaged in séance writing. Neither, we might add, was Darwin. In his *Biographia Literaria* Coleridge remarks of poetic endeavor that much may be gleaned from travels, books, natural history, and that all may be acquired as part of the writer's trade but that these cannot substitute for the ear of genius. He does not forget Lamb's dictum that "the true poet dreams being awake."[31] The artist dominates his subject.

Similarly it was from no "sunless sea" of memoryless dream that Darwin drew his own illumination. It was more like being led across the stepping stones of a brook into an enchanted land from which the first intruder, Edward Blyth, had leaped safely back to "reality." Darwin, a genius like Coleridge with "Reason at the Rudder," by contrast grasped immediately that he had come upon the long-sought magic which would bring order amongst all his idle facts and relate them in a rational pattern. He saw a vision for which he was prepared, but which he might never have glimpsed save for his perusal of Edward Blyth. The weary world traveler had had to come all the way back to London to find his secret in an unread magazine.

The widespread popularity of the "unconscious" theory concerning Charles Darwin can readily be explained by the fact that a cult of hero worship has developed about the great biologist, such as frequently happens to a prominent innovator in any field.[32] Darlington, the British geneticist, has commented ironically: "Among scientists there is a natural feeling that one of the greatest of our figures should not be dissected, at least by one of us."[33] In the face of evidence that Darwin made unacknowledged use of material from Blyth, the theory of the unconscious is the easiest, most polite way of evading the exploration of a delicate subject. Numerous naturalists who would never treat contemporaries so gently under similar circumstances are eager to make a "sleep-walker" of a scientist whose letters and notes are models of persistent conscious inquiry upon a great range of subject matter.

The paleontologist George Gaylord Simpson, referring to Darwin's statement in his autobiography that he "never happened to come across a single one [naturalist] who seemed to doubt about the permanence of species," and Darwin's belief that he owed no debt to his predecessors, said: "These are extraordinary statements. They cannot be literally true, yet Darwin cannot be consciously lying, and he may therefore be judged unconsciously misleading, naive, forgetful, or all three."[34]

Nora Barlow has also used the "unconscious" theory to explain her grandfather's denial that the subject of evolution was in the air. Doubtless Darwin's isolation at Down kept him from being aware of opinions from workers in other fields than his own, said Lady Barlow, "so that he unconsciously overlooked indications that belief in the permanence of species was waning."[35] Nevertheless some of the very journals he consulted contained references to the evolutionary hypothesis.

As opposed to the theory of the unconscious, it strikes one that Darwin was, in general, a keenly alert, conscious thinker, and he was so characterized by his associate, Thomas Huxley.[36] It is strange that in Darwin's *The Descent of Man* and *Variation of Animals and Plants Under Domestication*[37] all factual material drawn from Blyth was carefully listed but the two papers of Blyth concerning natural selection should be quietly ignored. It is difficult to accept this as mere coincidence. In *Variation* a footnote refers to the same volume of *The Magazine of Natural History* of 1835 in which Blyth's first paper on natural selection appeared.[38] Also a footnote in *Variation* contains the somewhat cryptic and unenlightening statement, "Mr. Blyth has freely communicated to me his stores of knowledge on this and all other related subjects."[39] There is no possibility of doubt that Darwin used and studied *The Magazine of Natural History* in which Blyth's papers appeared.

Another odd circumstance has recently been brought to light by Gavin de Beer, even though he has refrained from any comment as to its potential significance. I refer to the recent disclosure that a number of pages are missing from Darwin's

First Notebook on Transmutation of Species. The great importance of the first notebook in tracing Darwin's early thought has been stressed by de Beer.[40] Yet fifty pages are missing from this notebook, in which Darwin wrote on the first page: "All useful pages cut out. Dec. 7/1856/. (and again looked through April 21, 1873)."[41] Nothing was said about destroying the notes. As his son, Francis Darwin, pointed out in reminiscences of his father, Charles Darwin "felt the value of his notes, and had a horror of their destruction by fire. I remember, when some alarm of fire had happened, his begging me to be especially careful, adding very earnestly, that the rest of his life would be miserable if his notes and books were to be destroyed."[42]

De Beer, who reported in 1960 on these missing pages, said they had been searched for unsuccessfully in the Cambridge University Library, at Down House and the Royal College of Surgeons, and in the British Museum of Natural History. "The nature of their contents can only be surmised after a close study of the two hundred and thirty pages that remain," de Beer remarked, "and an estimate can be made of what is missing from the information and the argument."[43] Although there are some pages missing from the other notebooks, it is those from the first notebook that would seem to have the most bearing upon the origin of Darwin's theory, since it was begun in July 1837, before the date when he said he received his inspiration from Malthus. To reiterate my own words, I believe it significant that "Darwin opened his first notebook on the 'species question' in 1837. In January of that year Edward Blyth ventured the beginning of a second paper in which there is comment upon the principle of natural selection."[44] This comment, as we have seen, goes considerably beyond Blyth's first statement of 1835. It introduces, if briefly, the possibility of organic change. The name and work of Edward Blyth are not noted in the existing portion of the first notebook, although they do appear in the second.

"The idea of natural selection, so far as can be seen from the extant portions of the notebooks, seems to have occurred to Darwin as a combination of the effects on him of the facts of variation, adaptation, and extinction," observed de Beer.[45] Actually the missing fifty pages could have contained a great deal of information extending to Blyth's own views on these subjects. De Beer has avoided the suggestion that this fragmentary document may have contained more detailed references to Blyth's works. Since these pages compose the first part of the diary, their disappearance, taken with other evidence, cannot fail to hint of a genuinely "missing link" in the story of natural selection.

Much has been made by some of Darwin's defenders of his poor memory, though others have maintained it was prodigious. Huxley, who certainly knew him well, contended that Darwin had "a great memory."[46] Darwin himself remarked of his memory, that it "suffices to make me cautious by vaguely telling me that I have observed or read something opposed to the conclusion which I am drawing, or on the other hand in favour of it; *and after a time I can generally recollect where to search for my authority.*"[47] (Italics L. E.)

It is also true, however, that Darwin did not have to depend upon memory, as he was a remarkably methodical man in his work. In discussing his work habits he mentioned the fact that since in several of his books he had used extensively facts observed by others, he kept

> from thirty to forty large portfolios, in cabinets with labeled shelves, into which I can at once put a detached reference or memorandum. I have bought many books, and at their ends I make an index of all the facts that concern my work; or, if the book is not my own, write out a separate abstract, and of such abstracts I have a large drawer full. Before beginning on any subject I look to all the short indexes and make a general and classified index, and by taking the one or more proper portfolios I have all the information collected during my life ready for use.[48]

One statement of Darwin's, to which I have previously referred, is curiously revelatory to the student of character. In regard to an incidental matter of priority upon another biological matter, he wrote in his autobiography: "It is clear that I failed to impress my readers; and he who succeeds in doing so deserves, in my opinion, all the credit."[49] There is a strange indifference to historical priority here. Was Charles Darwin engaged in psychologically justifying a philosophy which permitted him to dismiss forerunners from whom he had drawn inspiration—men like his friend, "poor Blyth," who "failed to impress" and therefore deserved no recognition from the world?[50] One is forced to reflect upon this possibility, which has even been seized upon and brought forward by later writers as a justification of Darwin's attitude toward his predecessors.

There will always be an ineluctable mystery concerning the origin of the theory of natural selection, just as there will always be a shadowy web surrounding the real Charles Darwin, a web unseen but as real as the black cape in which we see him enveloped in a photograph taken of him on the verandah at Down at the age of seventy-two. One of Darwin's most ardent supporters, George Gaylord Simpson, states with perceptive acuteness: "The mystery persists. The man is not really explained, his inner adventures are not fully revealed in his own autobiography, in the family biography by Francis Darwin, or in the many other biographical sketches and books. There will always be something hidden, as there is in every life. . . ."[51]

One conclusion seems inescapable, though it does not solve the mystery—that there is a remarkable similarity between Coleridge, the "library cormorant," as he chose to describe himself, and Darwin, the reticent observer of nature and devourer of books. Each man had his catalyzer, whose existence it has taken over a century to unearth.

1. CHARLES DARWIN. This photograph of the great naturalist and commonly named discoverer of natural selection was taken in 1854, some years before the appearance of the *Origin of Species*.

2. ROBERT FITZROY *(opposite)* . The Captain of the *Beagle* was one of the important figures in Darwin's life. Darwin sailed as naturalist on the *Beagle* on one of the nineteenth century's most celebrated voyages (1831–1836).

3. DARWIN'S HOUSE AT DOWN. Shortly after his voyage around the world, Darwin isolated himself and his family in a little village in Kent, where he developed his theory.

4. ERASMUS DARWIN. Charles Darwin did not compose the theory of evolution out of thin air. All of the elements which were to enter into it were being widely discussed during his college years. His own grandfather, Erasmus Darwin, had anticipated views on evolution later expounded by Lamarck.

5. JEAN BAPTISTE DE LAMARCK. The French naturalist was another forerunner of Darwin in evolutionary thought. He proposed a theory that changes in environment cause changes in structure of animals and plants and that such acquired characters are transmitted to offspring.

6. CHARLES LYELL. Regarded as the father of modern geology, he opened the way for the evolutionary point of view by demonstrating that the earth must be very old, thus making possible extremely slow organic change.

7. ROBERT CHAMBERS. The Scottish publisher wrote and brought out a modified version of Lamarckian evolution. The popularity of his *Vestiges of the Natural History of Creation* (1843–1846) helped to prepare the way for Darwin's theories.

8. ALFRED RUSSEL WALLACE *(opposite)*. While visiting the Malay Archipelago, the young English naturalist originated independently the theory of natural selection (1858), an account of which he sent to Darwin.

9. ALEXANDER VON HUMBOLDT *(opposite)*. The German naturalist wrote *Personal Travels in South America*, the reading of which influenced both Wallace and Darwin.

10. THOMAS MALTHUS. The English economist's *An Essay on the Principle of Population* (1798) profoundly influenced Wallace in formulating his theory.

11. JOSEPH HOOKER *(above, left)*. The botanist, together with Charles Lyell, arranged to have a short summary by Darwin accompany the reading of Wallace's paper before the Linnean Society. Thus the theory arrived at by the two men was announced at the same time.

12. THOMAS HUXLEY *(above, right)*. The English biologist was the foremost advocate in England of Darwin's theory of evolution and defended the naturalist in debate.

13. EDWARD BLYTH *(opposite)*. The mysterious Mr. X, one of the forgotten parents of a great classic, is seen here as he appeared when he visited Dublin in 1864.

14. CHARLES DARWIN. He was born in 1809 and lived until 1882. This photograph of him was taken on the verandah at Down in 1881, when he was 72. There will always be a shadowy web surrounding Darwin, unseen but as real as the black cape in which he is enveloped.

PART

The Documentary Evidence

The historical documents which establish Edward Blyth as not merely a Darwinian precursor but in fact a direct intellectual forebear in a phylogenetic line of descent are to be found in aging files of periodicals in grimy cellars of old libraries. Originally published in The Magazine of Natural History *in 1835, 1836 and 1837, the essays by the naturalist are reprinted here under abbreviated titles. They show that Blyth indeed belongs to the royal line.*

The Varieties of Animals

I

THE APPELLATION "variety" being very commonly misapplied to individuals of a species, which are merely undergoing a regular natural change, either progressing from youth to maturity, or gradually shifting, according to fixed laws, their colors with the seasons, I conceive that it will be useful to some, to point out a few of the less generally known changes which naturally take place in various British animals; some few of which appear to have been hitherto overlooked, and others to have been described incorrectly.

The term "variety" is understood to signify a departure from the acknowledged type of a species, either in structure, in size, or in color; but is vague in the degree of being alike used to denote the slightest individual variation, and the most dissimilar breeds which have originated from one common stock. The term is, however, quite inapplicable to an animal in any state of periodical change natural to the species to which it belongs.

Varieties require some classification; and though I feel myself hardly adequate to the task, I shall here propose to arrange them under four principal heads; in the hope that this endeavor will induce some naturalists, more competent than myself, to follow out this intricate and complicated subject, into all its details.

I would distinguish, then, what are called varieties, into *simple variations, acquired variations, breeds,* and *true varieties.* These appear, in general, sufficiently distinct, although the exact limits of each are sometimes very difficult to be assigned. Indeed, in many cases they only differ in degree, and in others they may be all combined in one individual. Moreover, the varieties of either class have a much greater tendency to produce varieties of another class, than the typical animals of a species have to produce any sort of variety.

I. Simple Variations. The first class, which I propose to style *simple or slight individual variations,* differs only in degree from the last, or *true varieties;* and consists of mere differences of color or of stature, unaccompanied by any remarkable structural deviation; also of slight individual peculiarities of any kind, which are more or less observable in all animals, whether wild or tame, and which, having a tendency to perpetuate themselves by generation, may, under particular circumstances, become the origin of true *breeds* (which constitute my third class of varieties), but which, in a state of nature, are generally lost in the course of two or three generations. Albinoes belong to this first division, and also the other numerous anomalies mentioned in VII, 589–591, 593–598.[1] These *simple variations* occur both in wild and in domestic animals, but are much more frequent in the latter, and are commonly observed in all *breeds* and *true varieties.*

Among the Mammalia, total or partial absence of color is always, I believe, continued through life; excepting, of course, the cases of mere seasonal change; and, in this class of animals generally, perfect albinoes are much more numerous than among birds. Perfect albinoes are peculiar to warm-blooded animals, and in them there is a total deficiency of coloring matter in the rete mucosum, and, consequently, in the fur, and even the pigmentum nigrum of the eye is entirely wanting. In birds, these *perfect* albinoes are extremely rare, although several instances have been recorded in VII, 593–598. There are three sorts, however, of true permanent albinoes, which may be thus designated: (1) *Perfect Albinoes;* which are entirely white, and in which the eyes appear crimson, from the total want of coloring matter, rendering the minute blood vessels visible; (2) *Semi-Albinoes;* which are either white or of a pale color all over, and in which the irides are always paler than usual, and not unfrequently blue [I, 66, 178]; and (3) *Partial Albinoes;* which are partly of the natural color, but are more or less mottled *permanently* with white; and in which, if a white patch surrounds the eye, the pigmentum of that organ is commonly wanting. I have thus observed a rabbit, one eye of which was red, and the other dark hazel; but such instances are of very rare occurrence, although (and it is a curious fact) rabbits are often seen wholly white, with the exception of a small patch around each eye; which organ, consequently, is of the usual dark color. Albinoes, when paired together, as is well known, produce chiefly albino offspring, and a *breed* of them may thus be perpetuated; but, even in a domestic state, they not unfrequently produce young of the usual color; and, if paired with an ordinary individual, they sometimes produce partial albinoes, or semi-albinoes [I, 178], and occasionally, if the original color be brown (as in the case of mice or rabbits), a black, sandy, or slate-colored offspring, or an individual with one of these colors more or less varied with white, is produced; but, in the majority of instances, the young wholly resemble one of their parents, and the *preponderance* is decidedly in favor of the natural hue.[2] The colored offspring of an albino, however, even if matched with another colored individual, has still a tendency to produce albinoes, and this fact has been noticed in the human species; but, as Mr. Lawrence observes on the subject (in his *Lectures on the Physiology, Zoology, and the Natural History of Man*), "the disposition to change is 'generally' exhausted in one individual, and the characters of the original stock return, unless the variety is kept up by the precaution above mentioned, of excluding from the breed all which have not the new characters.[3] Thus, when African albinoes intermix with the common race, the offspring generally is black," etc. These observations apply alike to all *simple* or individual *variations,* and to most other varieties, and afford one of many reasons why marked breeds are in a state of nature so rarely perpetuated. There is yet, however, before quitting this subject, another sort of albino to be considered, which, I believe, is peculiar to the feathered race, and which is not, like the others, permanent; these, therefore, I shall denominate *temporary* albinoes. Most of the pale, white, and pied varieties of birds, which are produced in a state of nature, are of this kind. A friend informs me that a perfectly white lark in his possession molted, and became of the ordinary hue. I lately shot a sparrow which was all over of a very pale brown, or cream color; it was molting, and some of the new feathers that were coming were of the usual color, and others were of a pure white: on the next molt, probably, no more white feathers would have appeared. Of a brood of young robins which frequented my garden, two were white, one partially so, and one of the usual mottled brown; these all molted into the ordinary color. I could add other instances to the list, especially amongst domestic poultry. But it does not hence follow that among wild birds there are *no* permanently white or pied varieties; or, in other words, no true partial and semi-albinoes. A blackbird with a white head has now inhabited a garden in this neighborhood for three successive years; and if the cupidity of collectors did not mark out every white or pied bird for destruction, I doubt not that I should have been able to have furnished some other similar instances of *permanent* variation.

II. Acquired Variations. The second class of varieties which I would designate thus, comprises the various changes which, in a single individual, or in the course of generations, are *gradually* brought about by the operation of known causes: such as the greater or less supply of *nutriment;* the influence of particular *sorts* of food; or, either of these combined with the various privations consequent upon *confinement;* which changes would as gradually and certainly disappear if these causes were removed.

Redundance or deficiency of nutriment affects chiefly the stature of animals. Those herbivorous quadrupeds which browse the scanty vegetation on mountains are invariably much smaller than their brethren which crop the luxuriant produce of the plains; and although the cattle usually kept in these different situations are of diverse breeds, yet either of the breeds gradually removed to the other's pasture would, in two or three generations, acquire many of the characters of the other, would increase or degenerate in size, according to the supply of nutritious food; though, in either case, they would most probably soon give birth to *true varieties* adapted to the change. In this instance, *temperature* appears only to exert a secondary influence. The Iceland breed of sheep, which feeds on the nutritious lichens of that island, is of large size; and, like the other ruminant animals which subsist on similar food, is remarkable for an extraordinary development of horns. Another example of *acquired variation,* dependent solely on the supply of nutriment, may be observed in the deciduous horns of the deer family, which are well known to be large or small according to the quality of their food. That *temperature* also does exert an influence greater or less, according to the species of animal, is very evidently shown in the case of the donkey, of which there are no breeds, nor true varieties, and but very few simple variations [VII, 590]: this animal is every where found large or small, according to the *climate* it inhabits.[4]

The influence of particular *sorts* of food may be exemplified by the well-known property of madder *(Rubia tinctorum),* which colors the secretions, and tinges even the bones of the animals which feed on it of a blood-red color; and, as another familiar instance, may be cited the fact, equally well known, of bullfinches, and one or two other small birds, becoming wholly black when fed entirely on hempseed. I have known, however, this change to take place in a bird (the small aberdevatt finch, so common in the shops), which had been wholly fed on canary seed; yet this by no means invalidates the fact, so often observed, of its being very frequently brought about by the direct influence of the former diet. In several instances which have fallen under my own observation, feeding only on hempseed has invariably superinduced the change.[5]

The most remarkable of acquired variations are those brought about in animals in a state of confinement or domestication: in which case an animal is supplied regularly with abundance of very nutritious, though often unnatural, food, without the trouble and exertion of having to seek for it, and it becomes, in consequence, bulky and lazy, and in a few generations often very large; while the muscles of the organs of locomotion, from being but little called into action, become rigid and comparatively powerless, or are not developed to their full size. The common domestic breeds of the rabbit, ferret, guinea-pig, turkey, goose, and duck, are thus probably only acquired variations, which, from the causes above-mentioned, have in the course of generations, become much larger and heavier (excepting, however, in the case of the turkey) than their wild prototypes, and less fitted for locomotion; but which, if turned loose into their natural haunts, would most probably return, in a very few generations, to the form, size, and degree of locomotive ability proper to the species when naturally conditioned.[6] The crested varieties of domestic geese and ducks, and the hook-billed variety of the latter, are, however, in all probability, *true varieties;* and what are called "lob-eared" rabbits may be either a *true variety,* or a *breed.* The various slight diversities, which I call *simple variations,* are very common in the present class of varieties; and there is also in them a great tendency to produce what I call *true varieties,* as well as those slighter deviations, which, by particular management, may be increased into the sort of variety I denominate *breeds.*

III. Breeds are my third class of varieties; and though these may possibly be sometimes formed by accidental isolation in a state of nature, yet they are, for the most part, artificially brought about by the *direct* agency of *man.*[7] It is a general law of nature for all creatures to propagate the like of themselves: and this extends even to the most trivial minutiae, to the slightest individual peculiarities; and thus, among ourselves, we see a family likeness transmitted from generation to generation. When two animals are matched together, each remarkable for a certain given peculiarity, no matter how trivial, there is also a decided tendency in nature for that peculiarity to *increase;* and if the produce of these animals be set apart, and only those in which the same peculiarity is most apparent, be selected to breed from, the next generation will possess it in a still *more* remarkable degree; and so on, till at length the variety I designate a *breed,* is formed, which may be very unlike the original type.

The examples of this class of varieties must be too obvious to need specification: many of the varieties of cattle, and, in all probability, the greater number of those of domestic pigeons, have been generally brought about in this manner. It is worthy of remark, however, that the original and typical form of an animal is in great measure kept up by the same identical means by which a true *breed* is produced. The original form of a species is *unquestionably* better adapted to its *natural* habits than any modification of that form; and, as the sexual passions excite to rivalry and conflict, and the stronger must always prevail over the weaker, the latter, in a state of nature, is allowed but few opportunities of continuing its race. In a large herd of cattle, the strongest bull drives from him all the younger and weaker individuals of his own sex, and remains sole master of the herd; so that all the young which are produced must have had their origin from one which possessed the maximum of power and physical strength; and which, consequently, in the struggle for existence, was the best able to maintain his ground, and defend himself from every enemy.

In like manner, among animals which procure their food by means of their agility, strength, or delicacy of sense, the one best organized must always obtain the greatest quantity; and must, therefore, become physically the strongest, and be thus enabled, by routing its opponents, to transmit its superior qualities to a greater number of offspring. The same law, therefore, which was intended by Providence to keep up the typical qualities of a species, can be easily converted by man into a means of raising different varieties; but it is also clear that, if man did not keep up these breeds by regulating the sexual intercourse, they would all naturally soon revert to the original type. Farther, it is only on this principle that we can satisfactorily account for the degenerating effects said to be produced by the much-censured practice of "breeding in and in."[8] There would almost seem, in some species, to be a tendency, in every separate family, to some particular kind of deviation; which is only counteracted by the various crossings which, in a state of nature, must take place, and by the above-mentioned law, which causes each race to be chiefly propagated by the most typical and perfect individuals.

IV. True Varieties. The last of these divisions to which I more peculiarly restrict the term *variety,* consists of what are, in fact a kind of deformities, or monstrous births, the peculiarities of which, from reasons already mentioned, would very rarely, if ever, be perpetuated in a state of nature; but which, by man's agency, often become the origin of a new race. Such, for example, is the breed of sheep, now common in North America, and known by the name of *ancons* or *otter* sheep.[9] A ewe produced a male lamb of peculiar form, with a long body, and short and crooked limbs: the offspring of this animal, with ordinary females, was found sometimes to resemble the one parent, and sometimes the other; but did not usually blend the characters of each; and, in the cases of twins, the two lambs were often equally diverse with their parents. This variety was extensively propagated, in consequence of being less able to jump over fences than the ordinary breeds of sheep. The solidungular ["donkey-footed"] variety of swine, tailless cats, back-feathered, five-toed, and rumpless fowls, together with many sorts of dogs, and probably, also the race of fan-tailed pigeons, are other striking examples of *true varieties.*

The deviations of this kind do not appear to have any tendency to revert to the original form: this, most probably, could only be restored, in a direct manner, by the way in which the variety was first produced.

To this class may be also referred, with more than probability, some of the more remarkable varieties of the human species. With regard to color, we know that temperature exerts no *permanent* gradual influence whatever: white races remain unchanged at slight elevations within the tropics; and the natives of Boothia Felix are very dark; the swarthy inhabitants of Mauritania are a white race, and their sunburnt hue is merely an *acquired variation,* which is not transmissible by generation, and which does not extend to those parts which are not exposed to the sun. The coloring principle of black races is inherent in them, and is quite independent of external agency; is even darkest in some parts which are the least exposed, and *vice versâ.* The Ethiopian race is nowhere more black than in the vicinity of the Cape of Good Hope, where the crops are sometimes injured by the winter's frost. Strangely enough, this invariableness of color constitutes about, perhaps, the most fixed character of these races.

There is one fact, however, here to be observed, which is very well worthy of attention; and this is, that colored varieties appear to have been chiefly produced in hot countries; which seems almost to induce the conclusion that they were originally efforts of nature, to enable the skin to withstand the scorching produced by exposure to the burning rays of a tropical sun.[10] How far the structural peculiarities of the Negro and other races may not, in some cases, be the effects of *breed,* it would be impossible, perhaps, now to ascertain, and would be worse than presumption, in a novice like myself, to try to determine. Wherever a black individual was produced, especially among rude nations, if the breed was continued at all, the natural aversion it would certainly inspire would soon cause it to become isolated, and, before long, would, most probably, compel the race to seek for refuge in emigration. That no example, however, of the first production of a black variety has been recorded, may be ascribed to various causes; it may have only taken place once since the creation of the human race, and that once in a horde of tropical barbarians remote from the then centers of comparative civilization, where no sort of record would have been preserved. But it is highly probable that analogous-born varieties may have given rise to the Mongolian, Malay, and certain others of the more diverse races of mankind; nay, we may even suppose that, in some cases, the difference, in the first instance, was much *greater,* and was considerably modified by the intermixture which must have taken place in the first generations.

The mixed offspring of two different varieties of man thus generally blends the characters of each; though instances are not wanting of its *entirely* resembling (like the mixed produce of an ancon sheep) either one or the other of its parents; but in this case (as in the albino) the perfect characters of the other parent frequently show themselves in the next generation. I am entering, however, into a wide field, already well trodden by many philosophers; and the subject is already probably pretty well understood by the great majority of readers. Those who are not so familiar with it, will find it ably treated in various works; especially in Dr. Pritchard's work on man, and in the published *Lectures on the Natural History of Man,* by Lawrence: some sound and excellent remarks on *varieties* will also be found in the second volume of Lyell's *Principles of Geology.*

Still, however, it may not be impertinent to remark here, that, as in the brute creation, by a wise provision, the typical characters of a species are, in a state of nature, preserved by those individuals chiefly propagating, whose organization is the most perfect, and which, consequently, by their superior energy and physical powers, are enabled to vanquish and drive away the weak and sickly, so in the human race degeneration is, in great measure, prevented by the innate and natural preference which is always given to the most comely; and this is the principal and main reason why the varieties which are produced in savage tribes, must generally either become extinct in the first generation, or, if propagated, would most likely be left to themselves, and so become the origin of a new race; and in this we see an adequate cause for the obscurity in which the origin of different races is involved. In a civilized state of society there are other inducements, besides personal attractions, and a new variety in this case, unless very *outré* indeed, would be gradually merged, and in a few generations would disappear entirely by intermixture with the common race. The inferior animals appear not to have the slightest predilection for superior personal appearance; the most dissimilar varieties of the same species mix as freely and readily together as the most typical individuals; the most powerful alone becomes the favorite. Instances of this are not rare in the breeds of dogs.

The above is confessedly a hasty and imperfect sketch, a mere approximation towards an apt classification of "varieties"; but if it chance to meet the eye, and be fortunate enough to engage the attention, of any experienced naturalist, who shall think it worth his while to follow up the subject, and produce a better arrangement of these diversities, my object in indicting the present article will be amply recompensed.

Here, however, I may observe, that the classification I have proposed for specific deviations in the animal creation, is equally applicable to those of the vegetable. The "varieties" in both are strictkly analogous.

II

I come now to the second division of my subject, which is to point out some periodical and other changes of appearance, which naturally take place in various British animals, and which do not constitute varieties. Among our native Mammalia, I know of three principal modes whereby a change of color is brought about; namely, an actual shedding of the coat; a partial shedding of the coat; and an actual change of color in the coat itself.

1. As an example of change of appearance produced by actual shedding of the coat, may be instanced the fallow deer *(Cervus dama),* whose white spots disappear with the annual casting of its coat in autumn.

2. Partial shedding of the coat takes place in those animals which acquire in autumn a covering of two different kinds: one long, downy, and warm, which is shed in spring; the other short and glossy, which is retained. This change of appearance is exemplified in the common water shrew *(Sorex fodiens),* the short summer coat of which is much blacker than the longer downy covering which conceals this in winter. In this little animal the additional winter coat is shed about the latter end of March, or beginning of April; and does not take place uniformly, but progressively, beginning on the head, and ceasing at the hinder extremities; and exhibiting in its progress, throughout, a well-defined line of separation. Animals which (as the British Mustelinae) have two sorts of fur, the *shorter* of which is the more warm and downy, do not undergo this change, but retain both sorts throughout the year. In these the young have only one kind, which is close and woolly; as is well exemplified in the

common polecat *(Putorius furc)*, the young of which are of a very uniform dark brown, and very unlike the old animals.

3. Actual change of color in the coat itself is exhibited in the appearance of the fallow deer's white spots in spring, and in the case of the mountain hare *(Lepus variabilis)*, which is in summer grey, adapted to the hue of the lichens on which it squats; and in winter white, hardly to be discerned upon the snow. The same change also takes place in the stoat or ermine *(Putorius ermineus)*, although this is doubted by Mr. Berry [VII, 591].[11] In mild winters, such as we have of late experienced in the South of England, but few of the stoats become white, and some of these not until the latter part of the season. The change takes place quickly, but not uniformly, the animal assuming for a short time a pied appearance; but I have not succeeded in ascertaining whether it is accelerated by sudden cold, as the animals are not always to be seen exactly when we want them. One perfectly changed, however, was seen in this neighborhood soon after the one or two days of very cold weather in the beginning of last October. In reference to Mr. Berry's communication, I may observe, that in many dozens of stoats which I have seen in summer, I have never yet seen a white one; whereas in winter, I have seen in the same neighborhoods a considerable number of white stoats. Where the climate is excessive, and the transitions of the seasons are more sudden, this change is much more likely to take place generally. In the fur countries, the ermine's change of hue is, I believe, most regular.

There has been, strangely enough, a difference of opinion among naturalists, as to whether these seasonal changes of color were intended by Providence as an adaptation to change of temperature, or as a means of preserving the various species from the observation of their foes, by adapting their hues to the color of the surface; against which latter opinion it has been plausibly enough argued, that "nature provides for the preyer as well as for the prey."[12] The fact is, they answer *both* purposes; and they are among those striking instances of *design*, which so clearly and forcibly attest the existence of an omniscient great First Cause. Experiment demonstrates the soundness of the first opinion; and sufficient proof can be adduced to show that the other is also sound. Some arctic species are white, which have no enemy to fear, as the polar bear, the gyrfalcon, the arctic eagle-owl, the snowy owl, and even the stoat; and therefore, in these, the whiteness can only be to preserve the temperature of their bodies [VI, 79]; but when we perceive that the color of nocturnal animals, and of those defenseless species whose habits lead them to be much exposed, especially to enemies from above, are *invariably* of the same color with their respective natural haunts, we can only presume that this is because they should not appear too conspicuous to their enemies. Thus, in the eloquent language of Mr. Mudie, who, however, advocates the first opinion, "the ptarmigan is lichen rock in summer, hoar frost in autumn, and snow in winter. Grouse are brown heather, black game are peat bank and shingle, and partridges are clods and withered stalks, all the year round."[13] So, also, on the Continent, the common red-legged partridge *(Erythropus vulgaris)* is of the color of the gravelly and sandy soils on which it is found. So, also, are the different larks, the common quail, the various snipes, and all the other ground squatters, of the hue of their peculiar localities. So, also, are the numerous small Grallatores which haunt the margin of the ocean, adapted to the color of the sand. So, also, are those sylvan birds, which quit the dense umbrage of healthy growing trees, to seek their food and expose themselves on bare trunks and leafless decaying branches, of the hue of their particular haunts. "So exquisitely are they fitted for their office," says Mr. Mudie, "that the several woodpeckers vary in tint with the general colors of the trees which they select. If it is an alternation of green moss, yellow lichen, and ruby tinted cups, with here and there a spot of black, then the green woodpecker comes in charge; but if it is the black and white lichens of the alpine forest or the harsh-juiced tree, then we may look for the spotted races upon the bark."[14] The wryneck is the color of the lichened branch; and the night swallow and the owls resemble their peculiar places of concealment. So, also, the gayer colors of nocturnal moths are always on the hinder wings.[15] and the anterior, which, when they rest, conceal these, are adapted to the hues of the various places where by day they are found: even the bright upper wings of the tiger moths *(Arctia caja,* and *A. villica)* are with difficulty recognized upon a lichened bank or paling.[16] It is curious, indeed, the resemblance which subsists between the colors of nocturnal birds and night Lepidoptera; the buff tip moth *(Pygaera bucephala)* thus reminds us of the barn owl *(Strix vulgaris);* and the goat moth *(Cossus ligniperda),* and a host of others, are similar in their tints to most of the Strigidae: in both cases they are doubtless intended for the same purpose, that of concealment. It would indeed be easy to extend this list of examples considerably further; but I shall only now mention the common hare, which, when in form, would hardly ever be seen were it not for its brilliant eye; if its eye were closed, which it probably was before its quick sense of hearing had warned it of our approach, it would almost always, perhaps, wholly escape our observation. This ever continued watchfulness must have given rise to the supposition, that the hare always sleeps with its eyes open.

Seeing, therefore, so many most striking adaptations of color to haunt, in cases where the concealment thus afforded can be the *only* purpose, I think it is not too much to infer, that the changes of color in many arctic animals were intended by Providence for the double purpose of preserving their bodily heat, and of enabling them to elude the observation of their enemies. Certain it is, that their *conspicuousness* would otherwise expose them to inevitable destruction. If I had here space, I could satisfactorily prove that the high-flying Falconidae can, in most cases, only perceive their prey when it is *moving;* just, as on the seashore, *we* can only distinguish sanderlings when they move. Small Mammalia which frequent open situations are rarely much abroad, except in the twilight; and ground-feeding birds are ever on the watch, and even the smaller kinds (as I have repeatedly observed) can perceive a hovering falcon *long* before it comes within the sphere of human vision; and they instantly flee to shelter, or they crouch, and lying motionless, so exactly resemble a portion of the surface, that even a hawk's eye cannot distinguish them. Why should the falcon race be endowed with such wonderful powers of enduring hunger and fatigue, if, as is said, at the elevations at which they soar, they can clearly distinguish every living object scattered over the wide expanse beneath them? It is only on such animals as are *off their guard* that they descend; or otherwise, food being so abundant, they would soon multiply to the extirpation of their prey; which, of course, would be very speedily followed by that of the preyer.

How beautifully do we thus perceive, as in a thousand other instances, the balance of nature preserved: and even here we see another reason why sickly or degenerate animals (those, I mean, which are less able to maintain the necessary vigilance) must soon disappear; and why the slightest deviation from the natural hue must generally prove fatal to the animal. How different, thus, are even *simple variations* from the seasonal changes of color which naturally take place! Properly followed up, this subject might lead to some highly interesting and important results. It certainly points to the conclusion, that every, even the slightest, tint and marking has some decided use, and is intimately connected with the habits and welfare of the animal; and it also furnishes a satisfactory reason, why closely allied animals (or, in other words, animals of very similar form and habits) should so very commonly nearly resemble each other in their colors and in the general character of their markings.

Seasonal and Other Changes in Birds

I

NUMEROUS AS ARE the writers in this department of zoology; assiduously as the study of birds is cultivated in all parts of the civilized world; and talented as are many of the naturalists and close observers who devote their more particular attention to this branch; it still appears to me, that the numerous and very diversified regular changes of plumage and general external appearance, observable in this interesting subclass of animals, have been hitherto very greatly and strangely overlooked, and that, in consequence, the many valuable physiological inferences deducible from their investigation have been quite lost to the purposes of science and of classification.

It is true that many naturalists have in so far attended to the mutations of plumage which some particular species undergo, as that they are able at once to recognize them in every livery they assume; but the exact ages, and seasons, of molting; the precise nature of the general, or only partial, change that is undergone, and the various accordances and dissimilarities observable between the changes of distinct species; the endless characters of agreement and difference, so important in pointing out affinities, in showing what apparently similar races could never be brought to hybridize together; would seem to have been passed over as unworthy of notice, as undeserving of a particular investigation.

The subject is both extensive and complicated, and involves a number of other recondite inquiries. I could have wished that some naturalist better qualified than myself had taken it in hand. For my own part, I have little time for practical observation; but, having long been in the habit of keeping a number of birds (chiefly the smaller kinds which occur in Britain) in a state of captivity, I have thus enjoyed some very favorable opportunities for making myself fully acquainted with the various changes that a great number of species undergo, both seasonally, and in their progress from youth to maturity and old age; and I have neglected no opportunity of studying those of other races, which circumstances may have variously chanced to place in my way.

It is to be remarked, then, that some species of birds (as, for example, the larks and starlings, the crows, the woodpeckers, and various others) molt the whole of their immature, or nestling, plumage the first year, including the wing and tail primaries; while a very few (as the bearded pinnock, *Calamophilus biarmicus,* and rose mufflin [*Parus caudatus* Linnaeus], *Mecistura rosea*) shed the primary feathers of the tail the first season, but not those of the wing: numerous other races (as all the modifications of the fringillidous and thrush types) molt their clothing plumage very soon after leaving the nest, and retain the primaries till the second autumn; the Falconidae, again, and some others, undergo no change whatever until that period. All those which I have as yet mentioned change their feathers only once in the year, towards the close of summer, immediately on the cessation of the duties towards their progeny: but there are various other tribes (as the wagtails and pipits, Motacillinae, and most of the aquatic races) which regularly undergo another general molting in the spring; though in no instance, that I am aware of, are the primary wing feathers shed more than once in the year: those of the tail, however, in some rare instances, are; and the different coverts, together with the secondary and tertiary wing feathers, in most, if not all, double-molting birds, are changed twice. In some migrative species (as the cuckoo, and most of the swallows), the young of the year do not change their plumage until the winter months; whereas the old birds molt in autumn; and in other birds, again (as in various ducks, [VIII, 544, 545]), two general changes of feather take place within the short period of about four months. Very many other similar diversities, of a more or less subordinate character, might be enumerated, if enough have not been already mentioned to show that a wide field for observation is here open to the practical ornithologist.

In like manner may analogous diversities be observed throughout the mammiferous subclass of vertebrate animals; thus, the squirrels and the shrews renew their covering twice in the year, and the rats and rabbits but once. The common squirrel's seasonal changes have never, that I am aware of, been remarked by any naturalist, though it is so common an inhabitant of our island: its summer coat is very different from that of winter, the fur being much coarser, more shining, and of a bright rufous color; while the ornamental tufts to the ears are wholly wanting: these grow in autumn, while the animal is renovating its coat, and continue usually till about the beginning of July, the time varying somewhat in different individuals. Their winter fur, besides being of a much finer quality and texture, is considerably longer, thicker and more glossy, and quite of a different hue from that of summer, inclining to greyish brown. The first young ones, too, which are produced very early in the season, push forth the winter garb, which, I believe, they then retain throughout the summer; whereas the second race of young ones, which, for the most part, make their appearance about midsummer, are first clad in the summer dress, which is exchanged, before they have become half grown, for that of winter. It is not improbable, also, that diversities of a like kind may obtain in the renewal of the scales of fishes.

What the definite purpose effected by very many of these peculiar and dissimilar changes may be, I confess myself utterly unable to say; nor can I suggest even a plausible hypothesis upon the subject. Why, for example, should the pipits *(Anthus)* shed their plumage twice in the year, and the larks *(Alauda)* but once? And why, also, should the latter change all their nestling primaries at the first molt, while the former retain theirs until the third (including the vernal) general renovation of plumage? It is easy enough to say, with Mr. Mudie, that, in the wagtails, and certain other species, the colors of the summer and winter dresses are each, in so far as they differ, more peculiarly adapted to the particular season of the year; but this is merely a concomitancy: in other words, this adaptation is not the purpose of the change; for we find that, in certain species which regularly molt twice in the year (as the tree pipit), the summer and winter plumage hardly differ; whilst, on the other hand, as complete an adaptation of color to season is effected in others (as the stone chat, and most of the Fringillidae), which molt in autumn only, by the wearing off of the extreme tips of the feathers; these in winter having covered and concealed another, and, in many instances, a very diverse, color beneath. By what reason can we ever hope to account for the curious fact, that the common drake, and also the pintailed and other teals, should molt their whole clothing plumage (including the tail) in summer, and then again in autumn? As Mr. Waterton has well remarked on the subject, "All speculation on the part of the ornithologist is utterly confounded; for there is not the smallest clue afforded him, by which he might be enabled to trace out the cause of the strange phenomenon. To Him alone, who has ordained the ostrich to remain on the earth, and allowed the bat to soar through the ethereal vault of heaven, is known why the drake, for a very short period of the year, should be so completely clothed in the raiment of the female, that it requires a very keen and penetrating eye to distinguish them." [VIII, 544.]

In one point of view, however, at least, a knowledge of these changes is of considerable practical use to the naturalist; for they not unfrequently point out at once, in doubtful cases, the most appropriate situation of a genus in a system, and thus assist him very greatly in his endeavors to fabricate a sound system of classification. Instances of this I shall not here advance, as it is necessary to say something first of what meaning I attach to that most hackneyed of all phrases, "natural system," concerning which it is more than probable that my views may very considerably, and perhaps essentially, differ, from those of many who may perchance honor them with a perusal.

Under this phrase, then, two very distinct kinds of relation are ordinarily blended together and confounded; viz., the adaptive relation of every organized production to the conditions under which it was appointed to exist, and the physiological relation subsisting between different species of more or less similar organization. These may be aptly designated the *adaptive system,* and the *physiological system;* the system of relative adaptation between the earth, its productions, and its inhabitants, and the system of agreements and differences between the organization of distinct races.

To illustrate the former of these is, perhaps, superfluous: it is the system by which alone the existence of one species is necessary to that of another, and which binds each race to its locality; where the presence of each is alike necessary to preserve the equilibrium of organic being around; and when circumstances have changed, and the necessity for its agency no longer remains, a whole race perishes, and the fragments of a skeleton in the solid rock perhaps alone proclaim that such had ever existed. It is the grand and beautiful, the sublime and comprehensive, system which pervades the universe, of which the sun and planets are but a portion, and which, to return to ornithology, is so well exemplified in the adaptation of the ptarmigan to the mountain top, and the mountain top to the habits of the ptarmigan; which suits the ostrich to the arid desert, the woodpecker to the forest, and the petrel to "the far sea wave." It is the majestic and admirable system by which all nature works so beautifully together, and to which all that our external senses reveal appertains. It is the system which, exquisite and intensely interesting in all its minutest details, is, if possible, even more so in its complicated relations; by which, by the *unity of design* pervading which, all is demonstrable to be the workmanship of One omnipotent and all-foreseeing Providence, under the beneficent dispensation of whom nought that ever exists or occurs stands isolated and alone, but all conduce and work admirably together for the benefit of the whole; by whose all-wise decree it is ordained, that, while the lofty and sterile mountain peak *attracts* the clouds, which in winter, in consequence, precipitate themselves upon it in the form of snow, it should thus *cause* itself to become clad in the hue of all others the most calculated to prevent its internal temperature from being farther reduced, and itself from thereby becoming an increased source of cold by radiation to all around; while, at the same time, the concretion of snow itself, instead of deluging the country round with superfluous moisture, is thus retained for a time upon the heights, not only to shelter the more tender organized productions of the mountain from severer cold, but also to furnish, by the action of the summer sun, a due supply of water, when needed, to the fountains and rills which irrigate and fertilize the more level country; there having done its part, to flow on to the mighty reservoirs of the ocean, again to arise in clouds, and to fulfill again its appointed rounds, with perpetual never ceasing energy, while the world endures.

Look around our world; behold the chain of love
Combining all below and all above.
See plastic Nature working to this end;
The single atoms each to other tend,
Attract, attracted to, the next in place
Form'd and impell'd its neighbor to embrace.
See matter next, with various life endued,
Press to one centre still, the general good.
See dying vegetables life sustain,
See life dissolving vegetate again:
All forms that perish other forms supply:
(By turns we catch the vital breath and die,)
Like bubbles on the sea of matter borne,
They rise, they break, and to that sea return.
Nothing is foreign: parts relate to whole;
One all-extending all-preserving Soul
Connects each being, greatest with the least;
Made beast in aid of man, and man of beast;[1]
All served, all serving; nothing stands alone:
The chain holds on, and, where it ends, unknown.

POPE'S *Essay on Man,* EPISTLE III

In this sense of the phrase only we trace what may be esteemed a suitable meaning to the term "natural system": this is the only system by which the wonders of creation are *naturally* arranged; this alone is the system which nature everywhere presents for our contemplation: but, admire it as we may, still this is not the system by which an extensive knowledge of *species* can be acquired, or which can be studied elsewhere than in the wilds.

Every species of organism, as must be obvious to all examining thinking persons, is framed upon a greater or smaller series of successively subordinate typical plans, upon each of which is organized a variety of different species, perfectly unconnected and distinct from each other, however some may resemble, even to minutiae, and which exhibit each typical or subtypical structure more or less modified, and, in the extremes, generally more or less *approximating* towards the extreme modifications of other plans of organization, in direct relation to the endless diversifications of the surface of the earth, to variety of climate, or to peculiar modes of procuring sustenance.

Thus far, I believe, all systematists agree.

I must venture, however, to differ from the majority of them, in opposing the prevalent notion, that the extreme modifications of diverse types blend and inosculate[2] by direct *affinity;* contending that, however closely these may apparently resemble, the most similar modifications of diverse types are not, in a physiological sense, more nearly related to each other than are the more characteristic examples of the same.

To this conclusion I was originally led by reflection on various interesting phenomena connected with the changes of plumage which take place in birds; having observed that, however importantly, to suit peculiarity in the mode of life, the general structure of very aberrant forms may be modified, so as to render it even doubtful upon which fundamental type they are organized there are, notwithstanding, certain constant characters, of less importance to the existence and welfare of the species, by which every typical standard may be easily traced to its ultimate ramifications; some of the most valuable of these characters, in the feathered race, being afforded by peculiarities in the mode of molting. To illustrate this, I may cursorily adduce the various finchlike Sturnidae (*Aglaius, Molothrus, Dolichonyx,* etc.); extreme modifications of the *Corvus* type; as are also, however unlike they unquestionably appear, the genera *Alauda,* and even *Ammodramus.* All these, I have ascertained either from direct observation, or from competent sources, shed the nestling primaries the first season, which is not the case with any modification of the fringillidous type, or of the dentirostral. If other characters be wanting, which point alike to the same conclusion, I may mention the constant presence of a craw, or enlargement of the esophagus, in all the Fringillidae, and its invariable absence in all, even the most aberrant, modifications of the *Corvus* type; all the latter, too, preserve the ambulatory mode of progression, which, in perfection, is not observable in any Fringillidae, not even *Plectrophanes.* Again, other characters of distinction between these two equivalent divisions are sufficiently visible in the general aspect of the bill, even where the extremes approximate: all the Fringillidae, for instance (to which I would restrict the appellation Conirostres), possess what may be strictly defined a *bruising,* or *compressing,* instrument: whereas the general character of the same organ in the other division is rather what may be aptly termed a *thrusting* one, intermediate in its structure between those of the Fringillidae and Dentirostres; in which last group the bill is modified into either a *snapping, holding,* or *tugging* instrument, as the case may be: sometimes all three, as in vireo.

However, to return to the proposition I was just advancing, that, *physiologically* speaking, there are no combinations of distinct types, no intermediate organisms, save those between a central type and its ultimate ramifications: the general structure may be intermediate, and, consequently, the situation a species holds in the *adaptive system,* the office which it may have to perform in the general economy of the universe; but the latter does not constitute *affinity;* neither, strictly speaking, is it *analogy;* therefore I must distinguish it by another term, *approximation.*

As I shall have occasion to make use of these words frequently, as I proceed, it will be necessary, before advancing further, to define the precise meaning which I attach to them, however much this may appear digressing from the subject more immediately in hand.

First, then, let us consider *affinity,* which, according to the views I hold, is inseparably connected with the doctrine of *types.*

All organized matter is, of course, intrinsically allied in its nature, as contradistinguished from that which is not organized; this, therefore, is the first, or, as some would rather say, the last, the ultimate, the slightest possible, degree of *affinity.* Next, we have a grand primary distribution of all organic matter into the animal and vegetable kingdoms; a division too obvious to be for a moment called in question, and universally allowed; admitted even, inconsistently enough, by those who hold that every natural assemblage of species, great or small, forms part of some quinary circle.[3] Now, I cannot but observe here, in passing, that, to any unbiased person, I should think that a due consideration of this first *binary* distribution must at once carry conviction to the mind, must be at once a most unanswerable argument against all *quinary* or similar doctrines; the which, of course, if based upon sound theory, would not only be found to hold good, but would be most obviously indicated by these primary and comprehensive assemblages of every created species. But, to return: here we have the animal type, and the vegetable type, diverse in structure, distinct even in chemical composition, insomuch that the *kingdom* to which any dubious production appertains may be decided by chemical analysis, even in a fossil, should but a very few particles of its primitive substance have been preserved. Say not, that the kingdoms blend at their ultimate extremities; for there are no better grounds for this supposition than those which led many, for a time, to advocate the spontaneous generation of Infusoria; extreme minuteness alone setting the limit to a definite partition. We must therefore admit, that there is a degree of physiological *affinity* between the most dissimilar animals, and also between the most dissimilar plants, which no animal or vegetable can possibly have for each other: species from the two kingdoms, however these may undoubtedly *approximate* at the extreme boundaries, *can* have no higher degree of *affinity* for each other than what they possess in common, as opposed to all unorganized matter; what further relations they may show are, therefore, totally distinct from *affinity.*

Leaving plants, we now enter upon the primary divisions of

the animal creation, the separate leading types, the distinct plans, upon one or other of which all animals are organized, and which cannot, any more than the last, be confounded, in any instance, one with another, however in particular cases these too may *approximate;* of which presently. Every vertebrate animal is, therefore, allied to every other vertebrate animal by what, to specify by numbers, may be expressed as three degrees of *affinity;* whereas it is physiologically related to every member of the Annulosa, and other invertebrate classes, by only two degrees, its affinity with plants being reckoned as one; the proportions of these numbers towards each other pretty accurately denoting the value of these degrees; two being double one, three exceeding by half two, etc. Animals of the same subclass, as different mammifers, or birds, or reptiles, are, of course, related to each other by four degrees of *affinity;* those of the same order by five, and so on; the number of these several degrees increasing in proportion to the number of subordinate successive types upon which different species are alike organized, and of which, successively, they are modifications, not *combinations* of different ones, in the last case any more than in the first. Every modification of every successive type is thus rudimentally different from the most approximate modifications of every other equivalent type, or superior type, to which it does not appertain; and this is the same conclusion to which I have been irresistibly led from consideration of various phenomena connected with the changes of plumage which take place in birds. As every species is perfectly and essentially distinct and separate from every other species, so, except in a retrograde direction, are the successive typical and subtypical plans upon which they are severally organized, however similar the latter may in some instances be, as are also the former. It is unnecessary to enter here upon any remarks on *hybrids,* as further elucidatory of the precise nature of *affinity:* it is well known that these can only be produced within a certain physiological range, and that their degree of fertility (paired with individuals of pure blood) is in proportion to the degree of *affinity* between the parent species.

By the term *approximation,* I must be understood to signify those modifications of particular types, which, adapted to intermediate modes of life, very commonly more or less resemble (in consequence of this adaptation) species which are organized on other and different types. I have already had occasion to mention certain extreme modifications of the corvine or omnivorous type of perching birds, which are close *approximations* towards the fringillidous type (as *Aglaius* and other finchlike Sturnidae, *Ammodramus,* and *Alauda*); the true *affinities,* however, of all which are at once shown by a reference to their molting. The hag, the lamprey, and the pride, are, thus, extreme *approximations* of the general vertebrate type of organization towards the class Mollusca. The *Ornithorynchus,* among mammifers, *approximates* very remarkably towards birds; but it exhibits less *analogy* with them, collectively, than many rodent species do. The pronghorned antelope is an *approximation* in its genus towards the Cervidae; but its *affinity* to the latter is not greater than in other antelopes. The frigate bird is an *approximation* towards the eagles; yet no one would consider it as organized upon the falcon type: so the *Pterocles* is an *approximation* towards the pigeons, and the Nicobar pigeon towards the Gallinidae; each being at once referable to its particular type, though in certain *adaptive* relations they are intermediate. The pipit genus is a most striking *approximation* of a very marked type (subordinate to the dentirostral) towards the larks; but its moltings at once intimate its true position in the system, however its general aspect might, at first sight, render this doubtful. It is by no means nearly allied by *affinity* to *Alauda;* and I will unhesitatingly venture to assert, that by no art could they be induced to unite to the production of a hybrid.

Analogy, in the most definite signification of the term, is well exemplified in the close resemblance between the mouth of the swift, and those of the larger high-flying insectivorous bats *(Vespertilio).* It is exhibited wherever species that are modifications of diverse types are organized to perform nearly the same part in the general economy of nature; which latter by no means necessarily implies *approximation;* as may be illustrated by adducing the vultures among birds, and the dog kind among quadrupeds, or certain of the Sphingidae from amongst insects, as compared to the Trochilidae of the feathered race. It is well exemplified by the deadly spring of the cats, as compared with that of the crushing serpents, and as somewhat contradistinguished from that of the saltatory spiders; all the energy of the body being, in the former cases, remarkably concentrated in a single spring, from which exhaustion follows, while in the latter case it is not. It is curiously shown by a fact related by Sir W. Jardine, of the European howler or eagle-owl *(Bubo europaeus),* in which the *analogy* of that genus to the cat family is even more strikingly indicated than by the very remarkable general resemblance in their external aspect. "This bird," observes Sir William, "evinces a great antipathy to dogs, and will perceive one at a considerable distance; nor is it possible to distract its attention so long as the animal remains in sight. When first perceived, the feathers are raised," etc., exactly as a cat raises her fur at sight of her natural enemy; though, in either case, it is difficult to say why they should be inimical. No doubt, however, the purpose, the reason for this antipathy, is the same in both instances, and it is for the naturalist to endeavor to find it out. The common pipit, a modification of the dentirostral type; and the Lapland snowfleck, one of the conirostral (as here limited); are in so far related to each other by *analogy,* as that they are both *approximations* towards the lark genus, an extreme modification of the omnivorous or corvine type; they are therefore related to each other by a certain analogy; to *Alauda,* by *approximation;* and to all the members of their respective separate groups, by an additional degree of *affinity* to what subsists between either of them and the others. *Affinity* and *analogy,* of course, coexist, as all organisms are, at least, related by what I have termed the first degree of the former; but the extent of the former does not necessarily affect that of the latter: vultures and dogs, for instance, are allied by three degrees of *affinity;* while the carrion beetles (Carabidae) are related to either by only two degrees: yet the *analogy* is as great in the one instance as in the other. Pure *analogy* may subsist with very trifling *approximation;* as is shown by the already cited case of the cats and serpents, or as may be exemplified by a hundred similar instances of corresponding groups existing in major divisions of diverse structure, in which, however marked the *analogy*, however similar the office they were destined to perform the degree of *approximation* is in many instances quite imperceptible.

Affinity, approximation, and *analogy,* may therefore be collectively defined as pertaining to the *physiological* relations subsisting between different species, as opposed to their *adaptive* relations; of which latter they are wholly independent: that is to say, different species, nearly allied by either of these physiological relations, exhibit no mutual, no relative adaptation towards each other's habits and structure; such as we observe in the huge claws of the anteater *(Myrmecophaga),* evidently furnished in direct relation to the habits of a particular group of insects, the mounds of which they are obviously intended to scrape open, while the tongue is as expressly modified to collect the aroused inhabitants, upon which alone the creature is fitted to feed, and upon the supply of which, therefore, as an existing species, its being altogether depends. *Adaptive* relations are, in general, even more obvious and striking in groups which are *physiologically* the most widely removed; as may be exemplified by adducing the bill of the crossbill, modified in direct relation to the seminiferous cones of the Coniferae; or the recurved bills of certain humming birds, to the bent tubes of the corollas of particular Bignoniaceae, etc. *Physiological* relations are all resolvable into mere *resemblance;* because every species is essentially distinct and separate from every other species; otherwise it would not be a *species,* but a *variety.* The most similar species, therefore, are only *allied* to each other in consequence of the close resemblance of their general organization; the degree of *affinity* being greater or less, according to the extent of that resemblance (according to the degree of their physiological, not their mere apparent, similitude); in short, according as they are more or less framed upon the same general or typical plan; which plans not only regulate the minutiae of structure in those species which are organized upon them, but, to a very considerable extent, even their colors and markings.

Of course, the observation here very naturally suggests itself, that, if the colors and markings of species have a definite use (which, in some instances, is sufficiently obvious even to our comprehension), then, we might reasonably expect to find that resemblance which is found to subsist between those of species whose habits are almost the same.

True; but, then, there are many trivialities observable in the marking of allied species, which can only be explained upon the principle that they are modifications of some particular general or typical plan, of markings, as well as of structure. Such is the pale line along the head of the newly discovered Dalmatian, *Regulus modestus* Gould, in place of the bright-colored coronal feathers of its different congeners; which is exactly analogous to the curious fact, that the apparent rudiments of dentition exist in the gums of the fetal toothless whales; sufficiently intimating that these latter animals are modifications merely of some general typical plan, of which one of the leading characters is to be furnished with teeth. So, also, might be adduced the tiny, soft, deflected spine situate at the bend of the wing of the common gallinule, in like manner indicating that this species, also, is framed upon some particular plan of structure, the more characteristic examples of which have spurred wings, as we find to be the case in the allied genus *Parra.* In all the species organized upon any given type, we may always look for some trivial resemblances of this kind; we may always expect to find some traces of any particular structure or markings, which are observable in those typical forms of which the others are but modifications; the probability of this, of course, increasing with the number of degrees of *affinity;* and it is not unusual, too, to find colors or markings, which, in typical forms are scarcely discernible, developed, as it were, in particular modifications of those forms, to a considerable extent: yet, in the most approximate modifications of diverse subtypes of one general type, we only find such trivial resemblances of this kind as may be directly traced up to the typical standard from which they both diverge; whatever other marks of similitude these may show being obviously analogous adaptations, rather, to similarities of habit, unaccompanied by those trivial resemblances which imply physiological proximity. Thus, however closely, both in form and coloring, our common grey flycatcher *(Muscicapa grisola)* may *approximate* to some of the smaller Tyrannulae of North America, the mottled character of its nestling garb at once indicates that it is not framed on exactly the same series of successive types; in a word, that its relation towards these tyrannules must be considered as one of *approximation,* rather than of direct *affinity.* It would be easy, in like manner, to illustrate the preceding several positions; but the limits of the present disquisition will not permit of it.

It only now remains for me to apply the various facts which I have been endeavoring to establish; after which I shall commence a minute detail of observations on the molting of birds. That our systems of classification should be founded on the true *affinities* of species, rather than upon any arbitrary characters, is now, I believe, admitted on all hands to be the desideratum; and the true *principle* on which alone this can be effected is, as it appears to me, sufficiently obvious; though, from our present very imperfect acquaintance with existing species, it must necessarily be a long while yet before our arrangements can be considered at all final, if, indeed, we can ever hope them to assume that character.

The true physiological system is evidently one of irregular and indefinite *radiation,* and of reiterate divergence and ramification from a varying number of successively subordinate typical plans; often modified in the extremes, till the general aspect has become entirely changed, but still retaining, to the very ultimate limits, certain fixed and constant distinctive characters, by which the true affinities of species may be always known; the modifications of each successive type being always in direct relation to particular localities, or to peculiar modes of procuring sustenance; in short, to the particular circumstances under which a species was appointed to exist in the locality which it indigenously inhabits, where alone its presence forms part of the grand system of the universe, and tends to preserve the balance of organic being, and, removed whence (as is somewhere well remarked by Mudie), a plant or animal is little else than a "disjointed fragment."

Systematists, with few exceptions, err most grossly in imagining that allied species have been created in direct reference to each other (as members of a sort of cabinet system of even proportions) rather than to the localities they indigenously frequent, to the office each was ordained to fulfill in the universal, or *adaptive,* system. One would have supposed that the various facts which geology has brought to light would have sufficed to undeceive them in this particular. It cannot be too often repeated, that, upon whatever plan a species may be organized, its true relation (the reason for its existence at all) is solely connected with its indigenous locality: else, why should so many thousand species have ceased to be, the particular circumstances under which they were appointed to live no longer requiring their presence? To expect, indeed, for a single moment, that, in any isolated class or division of organisms, a perfect system of another kind could obtain, harmonizing in all points, and true in the detail to any *particular number,* appears to me (even supposing that none of the species were now extinct, and that we knew all that are at present existing), *prima facie,* a manifest illusion. Species are distributed over the earth, wherever the most scanty means of subsistence for them are to be found; and their organization is always beautifully and wonderfully adapted for obtaining it under whatever circumstances it may exist: just, therefore, as the surface varies, so do its productions and its inhabitants; and there is no locality, or apparently, even vegetable production, so peculiar, but species are found upon it especially organized to find their subsistence chiefly or wholly there. The very underground lake has its own peculiar inhabitants; for the wondorous *Proteus* there revels in regions of everlasting night: of course happy in its existence as the bird that cleaves the free air, or as the lion that exults in his conquering prowess. Ponder this well; and it is clear, that upon these grounds alone all *quinary* imaginings must at once fall to the ground.

The more deeply, indeed, I consider the quinary theory (now advocated by so many talented naturalists) in all its bearings, the less consistent does it appear to me with reason and common sense; the more thoroughly am I convinced of its utter fancifulness and misleading tendency. Nothing in this world is without its particular and definite *use,* which observation, in time, generally contrives to discover: but what utility could there be, what *purpose* could be effected, by separate and distinct races of beings, created obviously in direct relation to particular localities, being distributed into even groups of a limited number, like the celebrated groves of Blenheim, "nodding at each other"? If the quinary system be universal, as some would have, pervading all creation, how is it that the stars and planets do not revolve in groups of five? Or why even do not animals mostly produce their young by fives, or multiples of five? The absurdity is, indeed, too great to be dwelt on. If we examine, too, the writings of even the most eminent advocates of this strange theory, we continually meet (as might be expected) with divisions apparently made for mere dividing sake, that the requisite number of groups might be filled up; and, on the other hand, with examples equally glaring of the most dissimilar forms being brought under one general head, that the same particular number should not be exceeded. Thus, in Mr. Selby's in many respects very valuable and useful *British Ornithology,* while the closely allied linnets and siskins are placed in *separate subfamilies,* between the *types* of which no supergeneric character of the least importance can be described, we find the buntings actually arranged in a subfamily of which the larks are typical; and, in another division, of like value, among his Sylviadae, four genera *(Parus, Accentor, Setophaga,* and *Calamophilus)* grouped together, which have hardly a single character in unison that is not common to the whole Dentirostres, and which, certainly, are but very distantly allied. To adduce additional instances must be superflous: a system which can admit of such very arbitrary arrangements can have but a faint title indeed to be designated the "only natural one."

It is unnecessary now any longer to detain the attention of the reader by further prefatory observations; nor would it be worth while here to offer any remarks on the progress of plumification, the which might be better introduced as occasion may require; but I shall forthwith proceed to point out what I conceive to be of very great importance towards the classification of birds according to their true affinities, the different changes of plumage and appearance to which various groups of them are subject, confining myself, for the most part, to those upon which I can speak quite positively, from having myself had opportunities of witnessing them. On this inquiry there is, indeed, hardly any guide to go by, but direct personal observation; for though, in the books the greater number of these changes of appearance in the feathered race have been often mentioned, it is seldom that the precise manner in which they are brought about is stated; and the term "vernal molt" has been, in general, so very vaguely applied (sometimes indicating an actual shedding and renovation of the feathers themselves, and sometimes merely the seasonal wearing off of their winter edgings), that I have thought it best to decline altogether availing myself of their assistance. I may just premise, however, before commencing, that, independently of molting, there are two principal modes by which a great alteration in the appearance of the feathers of birds is, in some cases, gradually brought about; namely, a direct change of color in the feathers themselves, and the gradual shedding, in spring (as has already been spoken of), of their extreme tips, which are frequently of a different and more dingy color than that part of the feather which becomes exposed to view when these have disappeared. A familiar and beautiful illustration of both these changes is furnished by the breast plumage of a male of the common, or song, linnet *(Linaria cannabina).* The colored portion of these feathers, in winter, is of a brownish red; and they are tipped with deciduous dusky edgings. In the spring, the latter gradually wear off, and the dark maroon changes to a bright crimson.[4] The same plumage which the ptarmigan acquires in autumn becomes, in winter, white, and in spring gradually reassumes somewhat its former color, but a still deeper one.[5] Variations in general appearance, however, induced by a change of color in the feathers themselves, are of comparatively rather unusual occurrence.[6]

II

There are two modes of estimating the typical standard of a natural group of species. There are two distinct principles upon which, according as we desire to frame a system upon obvious and tangible characters, or upon the physiological relations, that is the true affinities, of species, we may arrive at very different conclusions as to which form is the more worthy to be considered the general type of the whole. I have said that it is not unusual to find certain characters, which, in typical forms, are comparatively little noticeable, carried out, in particular modifications of those general plans of structure, to a much greater extent; in exemplification of which may be adduced (as a familiar, though not, perhaps, the most striking instance) the great development of the bill laminae in the shoveler genus *(Spathulea);* also the perfection of the bill, as a groping instrument, and as a sentient organ, in the snipes and woodcocks; in consideration of which, many naturalists, esteeming these to be the most characteristic peculiarities of their respective major groups, have therefore adopted the above-named genera as the types of extensive natural families. Now, this may be very well in a confessedly artificial system; but, where *affinity* is to be considered the basis of classification, these forms will rather have to be arranged as ultimate modifications of their respective types, in a particular direction. They are neither of them *centers of radiation* (at least, to any extent), such as the form of *Anas boschas* undoubtedly is in the duck family; and such as the godwits *(Limosa)* at least approximate to be in the natural family to which the snipes appertain. *Corvus* and *Ardea* are good examples of thoroughly typical forms, which, modified in every possible way, radiate and ramify in every direction around; and so, also, is *Merula,* and that central division of the finch family to which the term *Coccothraustes* has been given. All of these graduate, through a series of species, into almost every form referable to their respective groups; and such must necessarily be the case with the more characteristic examples of every general plan of structure, of whatever value. Typical forms, in fact, as a leading rule, are merely those examples of each plan which are the least bound, as a matter of necessity, to particular localities; and we accordingly find them (I mean the *forms,* rather than species) to be of comparatively general distribution; whereas the more one of these plans is modified to suit any particular purpose, the more completely it is adapted to any peculiar sort of locality or mode of life: the *adaptation,* of course, implies a receding from the general, or central, type; and the species may therefore, in technical language, be termed *aberrant,* even though its deviation be a farther development of characters peculiar to its group.

It is clear that we must either admit this, or allow of a multiplicity of primary types to every natural family, to every group of species framed upon the same general or leading plan: the which must necessarily lead to such gross violations of *affinity* as the adoption of Phasianidae and Tetraonidae of the Quinarists as separate and independent natural groups, equivalent and equally distinct from each other, as are either of them from the two contiguously ranged families, Columbidae and Struthionidae; and this, too, while the very genera assumed to be typical of them, Tetrao and Phasianus, are allied so nearly as to hybridize together.

This is so interesting a subject, that a few additional remarks may be well devoted to its elucidation. Assuming a type to be merely the abstract plan upon which a certain number of species are organized, the said plan being variously more or less modified according to the purpose for which a species was designed, it certainly does not necessarily follow that organisms simply illustrative of the mere plan should have been created, seeing that all creatures are obviously framed in direct relation to their indigenous haunts, and not as mere counterparts of one another. At the same time, wherever an extensive array of species are organized upon one general plan of structure, there cannot but appear some tendency to converge to a general center; a tendency becoming more obvious as we recede from the extremes, whereupon there is usually a marked increase in the number of species exhibiting the same characters, till at length a sort of focus presents itself, as a central genus, the proper limits of which completely baffle the ingenuity of naturalists to define, inasmuch as the various species it comprises blend with, and continuously radiate into, the immediately subordinate divisions.

In illustration, it is sufficient to mention the already cited genera, *Corvus, Ardea, Merula,* and what should be *Fringilla,* but which is at present better known as *Coccothraustes.*

Take either of these divisions, and observe how difficult it is to define its (artificial) boundaries; how unbroken is the concatenation of species which links them with what are simply aberrant modifications of their structure, but which naturalists have been accustomed to consider as separate and distinct generic divisions. Let us, for a moment, consider *Merula.* Some naturalists try to separate the spotted-breasted thrushes from those in which the markings are less broken; and, unquestionably, taking the extremes, there is much diversity; but there is quite as much between the different spotted-breasted thrushes. In either case, however, where can the dividing line be drawn? The blackbird has, when young, a spotted breast; and, in fact, the characters of its nestling plumage alone forbid its alienation from the spotted thrushes. Where, indeed, can we trace the line of separation between *Merula* and *Philomela* even? And does not also the same form, in another of its various gradations, merge imperceptibly into *Petrocincla,* and thence into the different saxicoline genera, *Erythaca, Phoenicura,* and *Sialia?* one ousel (*Petrocincla,* or, rather, *Geocincla* Gould) being absolutely a large robin, another a great redstart, while a newly discovered species of *Sialia* has the markings, and many of the characters, of a *Petrocincla?* But it would be endless to follow *Merula* into all its diversified ramifications. I shall content myself with tracing the series into *Philomela,* which is at once conclusive as to the true affinities of the latter.

To be brief, then, we observe in the European song thrush a deviation from the gregarious character of its nearest British congeners, and an approximation to the style of marking in the transatlantic species. *M. mustelina* of North America is yet more solitary, and does not even associate to migrate; in this resembling *Philomela,* which its habits (as described by all who have observed them) accord with in almost every particular: still it retains a good deal of the true *Merula;* and it builds a plastered nest, like our thrushes. In *M. solitaria* the size decreases, the number of breast spots are diminished, the tarse is much lengthened (a character which commences in *M. mustelina*), the nest is constructed without plaster, and even the tail is rufous, as in the nightingales. *M. Wilsonii* has the very form of

Philomela, and is the smallest bird that ranks in *Merula:* its breast-spots are but very few, and these appearing as though more than half obliterated; its habits are exactly those of *Philomela,* and so is its nidification; and its bill hardly differs from that of our nightingale. The great nightingale of Eastern Europe has, according to Bechstein, an obscurely spotted breast, also a stronger bill than the common species; it is described to be more omnivorous in its diet, and, consequently, to be more hardy in a state of confinement: even its size implies an approach to the small *Merulae.* And, lastly, look to the nestling plumage of the song nightingale *(P. luscinia),* a character of no small importance in indicating the true affinities of birds, and we at once perceive its true station in the system, and how distinct it is from those forms with which (apparently from its mere size) it has been hitherto associated: it is, in fact, an ultimate modification of the type represented by *Merula.*

Let us now compare, for a moment, the extremes of the genus *Merula;* let us bring together the large mottled-backed thrushes of the East, and those diminutive solitary thrushes of the West. Does it seem proper that these should rank in the same minimum division? And yet how are they to be separated? How can the former be divided from those of the missel thrush form; the last-mentioned from the fieldfare group, the fieldfares from the merles, or from the congeries to which the song thrush belongs, which last we have seen to inosculate with the nightingales?[7] How, in like manner, can we divide the genera *Ardea* or *Corvus?*

It indeed appears that, in these very typical genera, there is a sort of clustering of species (if I may be allowed the phrase) about the center of radiation. In *Ardea,* and *Corvus* particularly, the central species become extremely difficult of determination; if, indeed, in some instances, the proximity is not even too close for detection. How nearly do some of the typical crows resemble! Upon the most scrupulous and minute comparison, C. L. Bonaparte was unable to discern the least difference between specimens of the European *Corvus corone* and the common crow of America; and he consequently infers their identity: yet who that attentively peruses the various descriptions of the latter, that considers well its marked gregarious habits, and the diversity of its note from that of our crow, can for a moment coincide with him in opinion? Look again, to the raven, that formerly was considered a bird of universal distribution, as was also the snipe. First, the African species must be detached, as Le Vaillant's description of it should long ago have indicated; for we find that, independently of the differences in the bill, and certain particulars connected with its plumage, the proportionate size of the sexes is the reverse of that of the European species, as Le Vaillant himself ascertained and published. Then the beautifully glossed raven of the Brazils is obviously distinct; also the raven of the South Sea Islands, and, there is reason to believe, that of India. And what if these can be proved to be separate species, by fixed and constant structural distinctions; do they not show how nearly species may resemble, and point to the almost inevitable conclusion, that, in some instances, there may possibly be no means whatever of knowing them apart?

How vastly important is this consideration when we contemplate the natural productions of America! Many years have now elapsed since the genius of Buffon suggested the capital proposition, that there is no absolute specific identity between any organism of the Eastern and Western continents, with the exception of those which inhabit very far to the north. All subsequent investigation has gone to prove the force and acumen of this sterling remark; and the number of species (exclusive of evident stragglers) supposed to be common to the two continents has been gradually diminishing, on more careful and exact comparison from that time to the present. In fact, I think we may now fairly venture to assume, putting aside stragglers, that those species alone are satisfactorily identical in the two continents which are distributed over the whole north of Asia, and may be looked for on the north-western coast of America. Very lately, the American scaup (pochard) was found, on comparison, to be distinct from that of Europe, although the difference almost wholly consists in the obliquity of its wing spectrum; a character which, however, proved to be fixed and constant. Had there not been this diversity, the two species would have been, of course, equally distinct: yet how should we have discriminated them apart? The barn owls of the two continents, which are now believed (and on good grounds) to be distinct, are even more similar.[8]

Equally close resemblances obtain in other departments of the zoology of Europe and North America, and particularly in the insect tribes: many butterflies, for instance (as several of the *Coliades*), from the opposite shores of the Atlantic, being only to be told apart by the slowly acquired practical ken of the entomologist. The natural productions of Japan, again, in many instances, present the most astonishing similitude to those of Europe; yet they exhibit characters which cannot be well reconciled with variation, however unimportant in themselves, because they are distinctions which climate or locality are not in the least likely to bring about. Besides, supposing the latter, we should not only expect to meet with specimens in every degree intermediate, but to find the same species equally flexible to circumstances in other places, which is not the case.

In ornithology, the jay and bullfinch of Japan may be selected from among numerous other instances; the former differing only from the European bird in the greater development of certain markings about the head, and the latter presenting no other difference than the much paler, or roseate, tint of its abdominal plumage. Taking a series of species, we have every grade of diversity, from the obviously distinct Japanese peafowl *(Pavo muticus),* to the mealy linnet, which apparently, differs in no respect from that of Europe. In a specimen of a pettychaps from the same locality, the only difference I could perceive from our common *Sylvia trochilus* on very minute inspection, consisted in a peculiar slight curve at the extremity of the upper mandible: still we know how nearly two British species of this genius resemble, and yet how very diverse are their notes. Perhaps the song of the Japanese pettychaps is dissimilar from that of either: at any rate, a dry skin is hardly sufficient on which to found a definite opinion.

Of course, all these various facts lead us to the important consideration of, What is a species? What constitutes specific distinction? To which the only rational reply appears to be (and even this is quite incapable of probation), Beings derived from a separate origin. For it appears that hybridism, after all, is but an uncertain guide, however satisfactory in particular cases; there being much reason to conclude, from a general survey of the facts recorded, that, as the degree of fertility in hybrids (paired with individuals of pure blood) varies according to the degree of proximity in the parent species, the possibility of mules being produced at all existing only within the sphere of a certain affinity; so, on the other hand, when the parent species approach so nearly as some that I have had occasion to mention, their mixed offspring would be almost equally prolific, hybrid with hybrid. This is, at least, stated of all the members of the genus *Bos;* and most naturalists concur in the opinion, that our common fowls are derived from the blending of a plurality of species. Certainly, if the analogy of plants can be admitted, the fact is in so far settled; for I know many hydrid plants which of themselves yield fertile seed in abundance: the mixed produce, for example, of the *Calceolaria purpurea* and *C. plantaginea;* the former a half-shrubby species, the latter herbaceous.[9] A variety of additional instances could be enumerated. Hybrid plants, however, are equally sterile with mule animals, if the parent species are not very closely allied.

It is to be hoped that, ere long, the experiments of the Zoological Society will have solved this curious and important problem. Already some highly interesting and complex hybrids have been obtained under their management.[10]

I have found it to be a very general opinion among naturalists, that specific diversity must of necessity be accompanied by some perceptible difference in the structure. To this I cannot accede, until I hear of a sufficient reason why it should be the case. We perceive every grade of approximation, till in the shrews, for instance, a slight diversity in the form of one of the back teeth comprehends all the difference. It is therefore presumed that, as so very trivial a deviation cannot be said to affect the animal's habits, for what purpose, then, does it exist, save to intimate the separateness of the species? But, surely, it will not be contended that species were created with a view that man should be able to distinguish them! Surely, differences were not imposed merely to facilitate the progress of human knowledge!

Is it not much more rational to conclude, that, as great differences in the structure import corresponding diversities in habit, so, by the same rule, minor differences also imply an equivalent diversity in degree? Let us, again, consider the American and European crows: here it would seem that specific diversity is unaccompanied by any structural deviation.

Of course, it is hardly necessary to hint the importance of these facts to geological enquirers: they intimate the excessive caution requisite ere we can venture to identify the fragments of an organism, when even existing species, in many instances, are not, probably, to be told apart. It must be admitted that they warrant a good deal of scepticism as to many of the identifications that have been assumed.

But to return now to the four typical genera, which have led to the above lengthy digression. I certainly do not conceive it necessary that there should be, in all instances, an unbroken gradation into the subordinate forms, similar to that from *Merula* into *Philomela;* for it is evident that the affinities of *Philomela,* and its relations to the thrush genus, would be the same, were there no intervening examples. Still it is reasonable to suppose that, generally speaking, such series would occur; not, however, for the mere abstract purposes of arrangement, but because there are grades in localities and modes of life. That there should be species variously modified upon any particular plan of structure, and that the deviation should be greater in one instance than in another, of course implies radiation from a general center; and the very circumstance that the same characters are more developed in one species than in another, necessarily also occasions a gradation in the particular direction, which may happen to be more or less regular, according as circumstances (*adaptive* relations) require. That there should be a slight break, for instance, in the series where the fringillidous type is modified into *Loxia,* is perfectly consistent with the nature of the deviation; but the true affinities of the crossbills are, nevertheless, equally recognizable, and the same may be said in other cases where the hiatus is much more considerable.

And here it will not be out of place to say a few words upon the terms *perfection, degradation,* and the like, as applied to natural productions. Let it be borne in mind, that, although every species is equally and wondrously perfect, even to the most trivial minutiae, in reference to the office for which it was designed, still, if we desire to cite an instance wherein the adaptation, if not more *perfect*, is, at least, more obviously remarkable and extraordinary than in another, it is to *aberrant* species, rather than to the central or typical exemplifications of a general plan of structure, that we must direct attention; inasmuch as the former exhibit those modifications of that plan, those adaptations to a peculiar mode of life, which are the most calculated to excite our wonder and admiration. Such forms as *Loxia* and *Recurvirostra* are sufficient illustrations of the position. There, perhaps, would be no objection to the word *degradation*, understood strictly in a classical sense; but, when we consider its popular, its *English* meaning, in which alone it will be apprehended by an extensive class of readers, no term should be more carefully avoided: the most *degraded* species absolutely happening to be those which are the most worthy our especial admiration.

The difficulties of classification arise from the necessary fact (obviously necessary when we consider the adaptive relations of species) of there being successive centers of radiation; the different modifications of a leading plan of structure radiating in their turn, and thereby constituting an irregular series of subordinate types, of every degree of value. Thus, the starling type is comprehended in the omnivorous or corvine plan of structure, and, in its turn, comprises others of less importance, upon all of which may be organized an indefinite number of species, diversely modified to suit a variety of localities, and often *approximating* in external appearance to species framed upon other general types of structure, wherever they are alike modified to perform the same office in the *adaptive* system: such *approximation*, however, by no means inducing an additional degree of physiological *affinity*.

Before concluding this, I must call attention to another point worthy of consideration. To recur again to the four typical genera we have all along been considering, and which, of course, it is most satisfactory to revert to in every instance, it appears that the central species, for the most part, exhibit a marked increase of size, being generally about the largest of those framed on their respective plans of structure. I do not say that this obtains in every instance, but still it is so general as to be quite worthy of attention; and the rationale of it appears simply to be, that, as typical forms are more adapted for general distribution, and better calculated for finding subsistence in a variety of localities, than those modifications of them which are organized expressly for peculiar places only, we must infer that an increase of stature would, as a general rule, be incompatible with the well-doing of aberrant races; or, to put it inversely, that beings of comparatively large size require to be less partial in their adaptations; that (their wants being greater) they should not be too much confined to particular places for the needful supply of food. However, this is a rule so broken into by exceptions, and so entirely dependent on the character of the particular adaptation, that, though obvious enough in the main, it is much more likely to meet with assent than demonstration. Certain it is, that, in very many groups, the largest species are among the most centrally typical. Witness, by way of example, the woodpeckers and the parrots.

In fabricating an arrangement according to the natural method (i.e., based on the true *affinities* of species), we cannot be too much impressed with the consideration that organisms must be ever regarded in their totality; that no one structural character can be expected to hold in all instances, however important in particular cases. We have only to consider the fact, that, in a natural group, it is but the same leading plan of structure which is so variously modified, each organ, in its turn, being adapted differently to diverse circumstances; and we perceive how valueless are the arbitrary characters of those who try to frame artificial systems. Even the dentition of the Mammalia, so paramount in the majority of cases, becomes quite a secondary means of distinction in the Marsupialia; and the structure of the bill in birds, so important and corresponding a character in most instances, yet loses almost all its value in the Certhiadae. Unquestionably, all the yoke-footed tribes are very nearly related by affinity; yet how discordant are they in the details of their structure! A single, and comparatively trivial, resemblance in the organization of the foot becomes, in this instance, a character of the very first importance.

Psychological Distinctions Between Man and Other Animals

I

THERE IS NOT, within the wide range of philosophical inquiry, a subject more intensely interesting to all who thirst for knowledge, than the precise nature of that important mental superiority which elevates the human being above the brute, and enables man alone to assume the sway wheresoever he plants his dwelling; and to induce changes in the constitution and adaptions of other species, which have no parallel where his interference is unknown.

I am led to offer a few remarks on this subject, by observing continually that the instinctive actions and resource of animals are attributed, most inconsiderately, to the habitual exercise of their reflective faculties; often where it is utterly and manifestly impossible for them to have observed facts whereon to base those inferences, which alone could have led them, by an inductive process, to adopt the course we find them to pursue. I am perfectly aware that the word "instinct," by not a few, is denounced as a mere cloak for ignorance, as a sort of loophole through which to escape from a rational explanation of phenomena; but, with all deference to those who advocate this over and above refined notion, I venture to maintain that it has a very definite signification, to express which no other term could be substituted: it implies an innate knowledge, which is not, like human wisdom, derived exclusively from observation and reflection, and to assign a secondary cause for which is clearly impossible; wherefore it savors rather, I conclude, of sophistry, to affect to be dissatisfied with any non-misleading expression, which is currently understood to denote it.

Place a juvenile chimpanzee in presence of one of its natural enemies; a python, or one of the larger Feles; and it "instinctively" recoils with dread. But does a human infant evince the like recognition? Here, then, is a fundamental distinction at the outset.

Not only, too, do brute animals (as remarked by White of Selborne) attempt, in their own defense, to use their natural weapons before these are developed, but they intuitively understand the mode of warfare resorted to by their brute opponents. They know, also, where the latter are most vulnerable, and likewise where their concealed weapons lie. Observe the deportment of a rat that is turned into a room with a ferret: see how artfully he guards his neck against the wall, instinctively knowing that there only will his enemy fix.[1] Notice, on the other hand, the wondrous accuracy with which the Mustelidae constantly wound the jugular vein of any bird or quadruped they attack. Witness a thrush that has captured a wasp, first squeezing out the venom from its abdomen, before it will swallow it. Or see a spider trying to shake off a wasp from its web, and, failing to do so, proceeding to cut it clean away. Can aught analogous be traced in the actions of inexperienced man? Whence, then, the acquired knowledge on which these animals could reason to act thus?

The distinction is, that, whereas the human race is compelled to derive the whole of its information through the medium of the senses, the brute is, on the contrary, supplied with an innate knowledge of whatever properties belong to all the natural objects around, which can in anywise affect its own interests or welfare; a sort of intimation, by the way, that all the inferior races pertain to some general comprehensive system, all the components of which have a mutual reciprocal bearing, and to which man only does not intuitively conform nor constitute a part of, except in so far as his bodily frame is of necessity subject to the common laws of matter and of organization.[2]

In every other species, each individual comes into the world replete with "instincts," which require no education for their development. A kitten reared by hand, or a bird raised from the nest, have the same language, the same leading habits, as the rest of their species, but little, if at all, modified by change of circumstances.[3] A kitten watches at a mouse-hole, though it has never seen a mouse; the squirrel proceeds by the easiest possible method to get at the kernel of its first nut, by invariably scraping, with its lower incisors, at the softer end, which it instinctively turns in its fore paws to the proper position; and the wasp, crawling forth from its pupa envelope, immediately commences feeding the neighboring larvae. The human infant, too, applies instinctively to the breast, like the young of all other mammalians; but, unlike those, it has to attain all its after-knowledge through the medium of its external senses. It looks to its nurses, and those about it, for information; and these are capable of so communicating their attainments, as very materially to assist the infant learner in its acquisition of knowledge. It is preposterous to assert the contrary, as has been done; or to pretend that it rests on the choice of the infant whether or not it will learn. Practically, it cannot help doing so; and it is equally monstrous to deny that human beings can so communicate the results of their experience, that, with what in addition is ever accumulating, each generation must necessarily rise in knowledge above the last. Unless the faculties were to be much deteriorated, it could not be otherwise. Who can pretend to deny the excessive influence of every generation upon that which immediately succeeds it; the influence both of precept and example? Imagine it possible for those of the present day to refuse to instruct; and what would then be the consequent condition of their offspring? Apply the same test to any other species of animal; and in what measure would the progeny be affected?

I wish not to defend the untenable doctrine, that the higher groups of animals do not individually profit by experience; nor to deny to them the capability of observation and reflection, whereby to modify, to a considerable extent, their instinctive conduct: neither do I assert that the human race is totally devoid of intuition, when I see the infant take naturally to the breast; when I perceive the force of the maternal attachment, and the ardor of the several passions: which latter, however, are, of course, but incentives to conduct common to both man and animals. In only the human species are the actions resulting from them unguided by intuitive knowledge. All I contend for is, that the ruling principle of human actions is essentially distinct from that which mainly actuates the brute creation, whence the general influence of the two is diverse in kind; and I mistake if I cannot establish the position.

The brief period that elapses before most animals are compelled to perform the part allotted to their species, precludes the possibility of their attaining sufficient information from external sources, and renders, therefore, the possession of a substitute for knowledge so obtained absolutely requisite. We have already seen that such a substitute is not wanting; but that all the knowledge necessary to insure their general welfare is intuitively conferred on the brute creation. Their various actions, in wild nature, are consequently based on this innate knowledge; which, being the same in every individual of the same species, in a natural state (that is, as completely uncontrolled by those peculiar changes of condition which man only, the exception of all other animals, can bring about), superinduces a normal uniformity of habit throughout the members of a species, which is rarely modified to any considerable extent by individual experience. Now, this uniformity is at variance with what reasoning from observation could possibly lead to; and, as it extends even to the resource of creatures of the same species, when driven to emergency, we have herein sufficient intimation that their wiles and stratagems, however consonant with what reasoning from observation might suggest, may nevertheless be purely instinctive, perfectly unalloyed with any wisdom resulting from experience.

To ascend from illustrations the least equivocal, let me here cite the nidification of the feathered tribes. Who, that considers the wonderful fact, that not only genera, but even species, of birds are for the most part distinctly indicated by their nests, can fail to recognize in this the operation of a principle essentially distinct from that which we understand by the word *reason?* which latter, in human beings, can of course, be only the result of observation and reflection.[4] We observe a similar marked uniformity in the fabrics and operations of all animals of identical species (man only exempted), endless examples of which will instantly recur to the reader in the insect tribes; and, if we consider the beaver, and others of the higher grades of animals which join their labors for mutual advantage, or are otherwise remarkable for what has thoughtlessly been deemed *their* ingenuity, the same truth will be found still to hold just as obviously apparent, and forbids us to attribute their proceedings to aught else than the dictates of intuition.

It is most commonly, however, in the resource of brute creatures, the wisdom they display in their expedients, that unreflecting persons fancy they discern the proofs of intellect identical with human; but, even here, this does not necessarily follow; for it is sufficient to refer to the cases which I commenced by detailing, to be assured that Providence has conferred instinctive wiles on animals as a resource against contingencies; the legitimate actions resulting from which according, perhaps, with what reason might dictate in like circumstances, we are therefore apt to conclude must necessarily have been induced by reasoning. To illustrate what I mean, let me adduce the simulation of death practiced by so many species, with intent to weaken the instinctive vigilance of their foes or prey. (That another animal, it may be remarked, should suffer itself to be thus duped, is most probably a result of acquired experience.) A cat has been seen to feign death, stretched on a grassplot, over which swallows were noticed sailing to and fro; and by this ruse to succeed in capturing one which heedlessly approached too near it. The fox has been known to personate a defunct carcass, when surprised in a henhouse; and it has even suffered itself to be carried out by the brush, and thrown on a dungheap, whereupon it instantly rose and took to its heels, to the astounding dismay of its human dupe. In like manner, this animal has submitted to be carried for more than a mile, swung over the shoulder, with its head hanging; till at length, probably getting a little weary of so uncomfortable a position, or perhaps *reasoning* that its instinctive stratagem had failed in its object, it has very speedily effected its release, by suddenly biting. The same animal has been known, when hunted, to crouch exposed upon a rock of nearly its own color, in the midst of a river, and so to evade detection by its pursuers; and we perpetually hear such cases brought forward as decisive proofs of its extreme sagacity. However, as regards the latter instance, will not a brood of newly hatched partridges instantly cower and squat motionless at sight of a foe?[5] and, as concerns the former, do we not find that many beetles, though just emerged from the pupa state, will simulate death every bit as cleverly as a fox or corncrake? Whence it surely follows that there can be no occasion to attribute the act to a reasoning process in the one animal, any more than in the other.

It would be unnecessary to enter here into any details on the obvious correlativeness of the dominant instincts of animals to the mode of life most congenial to their constitution, to remark on the mutual relations of habit and structure, and the exquisite adaptation of structure to locality. Hence, the natural habits of species of necessity bear reference to their indigenous haunts, as manifestly as their structural conformation. Thus, the elephant, which, like the other great Pachydermata, affects the vicinity of rivers and marshes, delights to relax its rigid hide in the stream; and afterwards covers it with a thick plastering of mud, probably to retard its too rapid desiccation: the which has been deemed an incontrovertible proof of its reasoning from observation. A young robin, however, the first time that it sees water, will, if it be not too deep, fearlessly plunge in and wash; and a young wren or lark will avail itself of the earliest opportunity to dust its feathers on the ground, the exact purport of which is not yet definitely understood. If, therefore, the latter be thus obviously instinctive, what reason have we to esteem the former otherwise? The uniformity of all these habits and propensities, in creatures of the same species, tends rather to intimate that in neither case are they the result of reasoning.

To infer reflection on the part of brutes, as many have not scrupled to do, as the motive for whatever in human actions could only be the result of reasoning, one would imagine to be too palpable a misapprehension to need serious consideration; yet some writers have gone so far as to attribute forethought to the dormouse, and other species which provide instinctively against the winter season. Perhaps it might be deemed a sufficient overthrow to this most shallow notion, to call in mind the migrative impulse; to inquire how the untaught cuckoo (raised by permanently resident foster parents) could reason that in another clime it should escape the rigors of a season that it had never experienced? But herein we have an additional principle involved, which will require a separate consideration. Proceed we, then, to examine into the presumed sagacity of those provident creatures, as the ant and harvest mouse, that habitually lay up a store for future need, and even provide against all possible injury from germination, by carefully nibbling out the corcule from each grain. Can any thing be more truly wonderful as a matter of instinct? All instincts are, indeed, equally wonderful. But it would certainly be even more extraordinary, if every member of these species were to be alike induced to pursue the same course by a process of reasoning. The following anecdotes will suffice to probe the intellect of these animals.

I have a tame squirrel, which, though regularly fed all its life from day to day, nevertheless displays the intuitive habit of its race, in always hoarding the superfluities of its food. Now, in its mode of effecting this, a superficial observer might fancy that he discerned a fair share of intelligence. Carrying a nut, for instance, in its mouth, it scrapes a hole with its fore paws in the litter at the bottom of its cage; and then, after depositing its burthen, scratches together the hay, or whatever it may be, over it, and pats it down with its paws. Moreover, it never fails to remember the spot, and will occasionally, when not wishing for food, examine the place to ascertain whether it be safe. But mark the sequel. I have repeatedly seen the same animal act precisely thus on the bare carpet, and upon a smooth mahogany table; yes, upon a table I have frequently seen it deposit its nut, give it a few quick pats down, and finally thus leave it wholly unconcealed.[6] The tits *(Pari)*, also, evince a like propensity of hiding food, one of their many resemblances to the Corvidae; and a tame marsh tit that I once possessed used habitually to drop the remainder of the almond, or piece of suet, that he had been picking, into the water-glass attached to the cage, although he never could thence reobtain it, and though his water was thus daily rendered turbid. I could narrate analogous instances without number.

Thus it plainly appears, that the instinct of each animal is adapted to its proper sphere; for the mode of life it was destined to pursue, and for that only. With this restriction (if such it can be called), it is in each case perfect. The actions of every creature uncontrolled and uninfluenced by man are invariably such as tend to the general welfare of its species; sometimes collectively, however, rather than individually (whence we hear of what have been termed "mistakes of instinct").[7] They evince superhuman wisdom, because it is innate, and, therefore instilled by an all-wise Creator. Indeed, the unpremeditated resource of animals, in cases of emergency, is oftentimes decidedly superior to that of man; and why? Because they need not experience for their guide, but are prompted to act aright by intuition.

In wild nature, this inborn knowledge of brutes thus abundantly sufficing for the attainment of all they require, there is, in consequence, but little to stimulate the exercise of their reflective faculties; and, accordingly, their general agency may be considered as passive, in effect analogous to the operation of the laws of matter. Even the "half-reasoning elephant," in the wild woods, is but a creature of unreflecting impulse, to an extent which wholly dissevers it from all community of mental attribute with the lowest grade of mankind. Witness the subdued tamed animal, which, traveling along its accustomed route, suddenly broke loose from its attendants, affrighted at the near yell of a tiger. At once its former submissiveness was forgotten: it rejoined the wild troops, and was again a free tenant of the jungle. Years rolled on, and it was retaken by the ordinary method. The sight of the stakes never sufficed to awaken its recollection; nor did the mode employed to secure it when entrapped. It was sullen and savage, and acted in nowise differently from its companions. By chance, however, its former keeper was present, who, after a while, recognized the animal. He gave the word of command, and it was instantly submissive;

all traces of its wild nature suddenly dissipated; its previous habits were forgotten; it was once more a reclaimed animal, and suffered itself to be led tractably to its place of confinement. Would a rational being have acted like this elephant?

II

Man only, by the habitual exercise of his reasoning powers, appears to be competent to trace effects to their remote causes; and is thereby enabled to recognize the existence of abstract laws, by assuming the guidance of which he can intentionally modify their operation, or, from observation, convert them to a means of accomplishing his various ends. It is thus he wields the principle of gravitation; and it is thus, from studying the inherent propensities and consequent habits of other animals, that, by judicious management, he contrives to subdue their instincts (as in the case of the elephant just mentioned), or to direct their force towards affecting other purposes than those for which they were more legitimately designed. But a more remarkable sequence of human interference is that by removing animals from their proper place in nature, and training them to novel modes of life, wherein the field for the exercise of their original instincts becomes much limited, their faculties of observation and reflection are, in consequence, brought more into play, in proportion as the former are rendered inefficient; till, at length, experience not unfrequently supersedes innate impulse as the main spring of their actions; more especially where they have become attached to a human master, and pass much of their time in his society. Yet even here the difference between man and brute is still manifest, in the transmission of acquired knowledge by generation, in the offspring inheriting as innate instincts the experience of their parents; so that the tendency of brutes is ever to become slaves to a certain amount of intuition, rather than beings dependent on their own intelligence.[8]

And here we recognize a fundamental principle of domestication, which is only gradually induced to any extent through a series of generations. Thus the elephant, though tamed, is not domesticated, for every individual is separately captured in a wild state; and we have seen that, when one of these returns to its proper haunts, its natural instincts having been only for a time subdued and rendered subservient (not eradicated), these have again become the incentives to its conduct, to the exclusion of those reasoning faculties which had only been excited into action under circumstances adverse to the efficient operation of the former. Far otherwise is what we observe in animals truly domesticated: witness the opposite conduct of even the newly hatched progeny of a wild and domestic duck, though incubated by the same bird. But here a question arises, that, as numerous instincts in domestic animals, which are now hereditary, are known to have been originally habits superinduced by man's agency, to what extent may not all the innate propensities and consequent habits of animals have originated in the acquired experience of their predecessors?

As with all other subjects, we must trace the series upward from its more simple phases. In the insect world, we discern the most complicated instincts; modes of procedure of which the consummate wisdom excites our admiration and amazement, and bearing reference to a future generation, in beings which are but creatures of an hour. Can it be supposed possible that the progenitors of these derived their habits from acquired experience, and transmitted them as innate instincts to their posterity? Here we must ascend to a higher source, which, being admitted, the marked uniformity, also, of the instinctive habits of all wild animals, before commented on, warrant us in concluding that these were from the first imprinted in their constitution, and may, therefore, be legitimately esteemed as forming part of the specific character.[9]

The tendency of human influence is everywhere to destroy whatever conduces not to man's enjoyment, as superfluous, and only cumbering the ground; but to secure, by every means the reasoning faculties can suggest, a due continuance and never-failing supply of all that tends to the gratification of our species. Brutes, on the contrary, evince indifference to whatever does not immediately concern them; and although, practically, their influence upon their prey is for the most part decidedly conservative, yet they individually continue to destroy without reflection, and endeavor not, by any forbearance, or plan resulting from reasoning, to insure the perpetuity of their provision. That the squirrel or jay should instinctively plant acorns is, of course, nothing whatever to the purpose: we have already tested the sagacity of the former animal; and we know that the latter, removed from its proper office in wild nature, will bury a bit of glass or clipping of tin as carefully as it does a seed.

It may be worth while to devote a few remarks to the consideration of the unintentional agency of brutes, towards not only preventing the over-increase of their prey, which would only lead to too much consumption of the food of the latter, and so bring about famine and consequent degeneration from insufficiency of nutriment, but likewise towards preserving the typical character of their prey in a more direct manner, by removing all that deviate from their normal or healthy condition, or which occur away from their proper and suitable locality, rather than those engaged in performing the office for which Providence designed them. In illustration, it will be sufficient to call attention to the principle on which many birds of prey are enabled to discern their quarry. When the tyrant of the air appears on wing, his dreaded form is instantly recognized by all whose ranks are thinned for his subsistence; and instinct prompts them to crouch motionless, like a portion of the surface, the tint of which all animals that inhabit open places ever resemble; so that he passes over, and fails to discriminate them, and seeks perchance in vain for a meal in the very midst of abundance; but should there happen to be an individual incapacitated by debility or sickness to maintain its wonted vigilance, or should its colors not accord sufficiently with that of the surface, as in the case of a variety, or of an animal pertaining to other and diverse haunts, that creature becomes, in consequence, a marked victim, and is sacrificed to appease the appetite of the destroyer: so profoundly wise are even the minor workings of the grand system; and thus do we perceive one of an endless multiplicity of causes which alike tend to limit the geographical range of species, and to maintain their pristine characters without blemish or decay to their remotest posterity.[10]

Thus it is that, however great may be the tendency of varieties to perpetuate themselves by generation, we do not find that they can maintain themselves in wild nature; nor do the causes which induce variation, beyond the occasional and very rare occurrence of an albino, prevail in those natural haunts of species to which their structural adaptations bind them. We have already noticed the anomalous influence of human interference in altering the innate instincts of the lower animals, thereby unfitting them to pursue the mode of life followed by their wild progenitors. It would be needless to amplify on the concomitant effects produced by domestication on the changes in the physical constitution and adaptations of the corporeal frame of animals, which oftentimes render them dependent on human assistance for continuous support, in the degree of their domesticity. Such changes are equally imposed on the vegetable world by cultivation; and they everywhere mark the progress of man, and exhibit in indisputable characters the diversity of his influence over the inferior ranks of creation, from any mutual and reciprocal influence observable among these latter.

I may cursorily allude to hybridism also, as a phenomenon, as far as can yet be shown, at least in animals, where fecundation cannot happen fortuitously, in every instance referable to human interference. As yet, I have failed to meet with a single satisfactory instance, wherein commixture of species could not be directly traced to man's agency, in superimposing a change on the constitution of the female parent. This is a subject of exceeding interest; and I am glad to avail myself of every occasion to endeavor to incite some to undertake its further investigation. There can be little doubt that certain of our domestic races, as the common fowl, are derived from a plurality of species, which, however, do not blend in wild nature; so that their union (assuming the hypothesis to be correct) may here, at least, be fairly ascribed to domestication. Still, when we consider that separate species (i. e., races not descended from a common stock) exhibit, as is well known, every grade of approximation, from obviously distinct to doubtfully identical, there appears, I think, sufficient reason at least to suspect that circumstances may sometimes combine to induce those nearest allied to commingle. That the mixed progeny, too, would in some instances be mutually fertile, I know in the case of the hybrid offspring of the *Anser cygnoides*, and the common goose; but, in birds generally, the converse nevertheless obtains, as is particularly instanced, I have learned, by the hybrid Fringillidae reared in confinement; and also the mule betwixt the common fowl and pheasant; the males of all which appear (from a variety of instances I have been fortunate in collecting) to have been incompetent to fecundate the eggs produced.[11] Perhaps the superior size, too, of these hybrids generally to that of either of their parent species may be explicable on the principle which occasions the large growth of capons. However, none of the species here alluded to are by any means so closely allied as many that are known to exist; and, therefore, as in the vegetable world the degree of fertility in hybrids is in the ratio of that of affinity between the parents, those derived from very approximate species being, apparently, quite as prolific as the pure race, analogy would lead us to infer that the same law holds in the animal creation. At present, we have no proof of it: and I may conclude the subject by observing that the cases of supposed union (apart from human influence) betwixt the carrion and hooded crows, so often insisted on, are inconclusive, inasmuch as it does not appear that the individuals were ever examined and compared, although black varieties of *Corvus cornix* have been several times known to occur. Indeed, I have myself examined a female specimen, on which were several black feathers intermingled with the ordinary ash color on the back.[12]

The agency of the human race has been likened to that of brutes, in the particular that, as man effects the destruction of one species, he necessarily advances the interests of another. How far he may permanently benefit the latter, might be discussed on principles that have been already expounded. More able writers, however, have put the inquiry whether man, by taking certain plants, for instance, under his protection, and greatly extending their natural range by cultivation, does not thereby unintentionally promote the welfare of the various species which subsist upon them. But, will it be argued that man, by vastly increasing the breed of sheep, is unconsciously laboring for the advantage of the wolves? As little can it be concluded, regarding the human race as progressive (in which it differs from all other species), that any race hostile to man's interests can be permanently benefited by his agency. The question, in short, resolves itself into one of time.

It has already been intimated that man is the only species that habitually destroys for other purposes than those of food. This leads me to a few remarks on the extinction of species. Without alluding, however, to the more direct agency of the human race towards accomplishing the destruction of every terrene species which conduces not in some way to our enjoyment, we will merely consider the natural causes which suffice to extirpate all other races, but are inadequate to effect the extinction of the human species. We have already seen that brute animals, in a state of nature, are merely beings of locality, whose agency tends to perpetuate the surrounding system of which they are members. It tends to do so, but is insufficient to effect this permanently; because, in the immensity of time important changes are brought about in every locality, by causes ever in operation, to which the faculties of the inferior animals are blind. They must, therefore, perish with their locality, unless distributed beyond the influence of the change; for their adaptations unfit them to contend for existence with the more legitimate habitants of diverse haunts, in proportion as they were suited to their former abode: and it must be necessary for creatures of instinct to be thus expressly organized in relation to their specific haunts, even to all the minutiae we perceive, in order to enable them to perform efficiently their destined office; which exquisite adaptation, however, cannot but of course disqualify them for maintaining their existence elsewhere. In man only we discover none of these partial adaptations, further than that he is intended to exist upon the ground; and the human race alone, in opposition to all other animals, takes cognizance of the progressive changes adverted to, and, from reflection, intentionally opposes obstacles to their course, or systematically endeavors to divert their energy. Man's agency, indeed, tends everywhere to alter, rather than to preserve, the indigenous features of a country; those features which natural causes combine to produce: in short, he strives against the united efforts of all other agents, insomuch that, wherever he appears, with his faculties at all developed, the aspect of the surface becomes changed: forests yield to his persevering labors; the marshes are drained, and converted into fertile lands; the very climate, accordingly, changes under his influence, which every way inclines to extirpate the indigenous products of the soil, or to reduce them, by domestication, to a condition subservient to the promotion of human interests. Does not, then, all this intimate that, even as a mundane being, man is no component of that reciprocal system to which all other species appertain? a system which for countless epochs prevailed ere the human race was summoned into being. His anomalous interference, therefore (for this work most aptly expresses the bearings of human influence upon that system), essentially differing from the uniform agency of all the rest, in not conducing to the *general* welfare, is thus shown to be in no way requisite to fill a gap in the vast system alluded to. All rather tends alike to indicate him a being of diverse, of higher destiny; designed, in the course of time, with the aid of physical causes ever in operation, and the presumed cessation of the creative energy, to revolutionize the entire surface of our planet. I will presently recur to this subject as regards marine productions. It is sufficiently evident, that, as the human species is bound to no description of locality, but alike inhabits the mountain and the plain, and is, by self-contrivance, enabled to endure the fervid heats of tropical climes, equally with the withering blasts of a polar winter, it is consequently proof against the undermining effects of those surface changes which suffice to effect the extermination of every other.[13] Its future removal, then, from this scene of existence, whenever that shall happen, will probably be brought about on another principle: how, it would be most useless to inquire. There is no reason, however, hence to anticipate that supernatural means must *necessarily* be resorted to, as a malignant disease might suffice to level all ranks in the dust. It is enough for my present purpose, to indicate in this the diversity of the human from all other species.

Some have argued the connection of man with the reciprocal system to which the inferior animals pertain, because, forsooth, he sometimes is annoyed by parasites. Without dwelling upon this topic, I may be allowed to say that it remains to be shown that any are peculiar to the human species. The certain fact, that different races of mankind are infested by distinct species, rather points to the conclusion, that, as the bed cimex can subsist and thrive away from human habitations, so also may even those species which abide on the person.[14]

III

I will now proceed to notice, and follow to its bearings, that mysterious impulse (if possible, even more incomprehensible than ordinary instinct) which guides a migrant animal to its destined haven; which also propels a bee towards its hive, and a pigeon homeward from one extremity of Europe to another; a principle, as will be shown, not wholly absent from the human constitution. The migrative restlessness displayed so forcibly by birds of passage, even when raised in confinement, and plentifully supplied with the nourishment they have been accustomed to (thus showing that insufficiency of food is not the predisposing cause, as is also intimated by the early departure of certain species from their summer haunts, after performing the duties of the season), is merely on a par with all other instinctive manifestations: and I may cursorily remark that, from much careful and attentive observation, I have determined, at least to my own satisfaction, that, as a general rule, it is in autumn mainly influenced by decline of temperature, and in spring by the breeding stimulus: the period of the incidence of which latter (though, undoubtedly, somewhat affected by temperature) is primarily dependent on specific peculiarity, and, secondarily, on constitutional vigor.[15] It is not the erratic impulse, however, so much as the guiding principle, that we have here to do with; that wonderful, most inexplicable principle, on which a diurnal migratory bird is not only, and by night, enabled to soar for even thousands of miles, over seas and continents, surmounting every obstacle, even lofty mountain ridges, in its course, impelled always in one unvarying direction, till it arrives at the proper winter quarters of its species; but, at the ensuing season, is also led back to its former abode, to the precise locality that it had previously set out from, having been known even to return to confinement. I conceive it unnecessary to detail observed instances of this astonishing fact, because, in the feathered race, it is well known to every student of natural history. It will be enough to mention, that I have an instance, on indisputable authority, of a lame redstart returning regularly for sixteen summers to the same garden.

Among mammalians, numerous instances have been recorded, resting on unexceptionable testimony, of animals returning straight to their accustomed haunts, over pastures and across streams they could not possibly have ever traversed before, and by a nearer and very different route from that by which they had been driven or carried. To these I will add the following, which occurred to the personal knowledge of my informant. A cat, from the center of an intricate and populous seaport town,[16] was shipped on board of a vessel bound for the Brazils; and, after performing the voyage to and fro, contrived to escape, on returning to its native port, and found its way, through several streets to its former domicile.

Mr. Jesse, in the third series of his *Gleanings*, has related a like anecdote of a reptile. Of a number of turtles, captured on Ascension Island, chanced to be an individual which, to use the technical phrase, had lost one of its *fins*. It was marked in the ordinary manner on the under shell, which marks are well known to be indelible. The vessel, on arriving in the Channel, was long detained by contrary winds, during which time a great mortality took place among the turtles; these dying one after another so fast, that it was at length resolved to cast what few remained of them, including the lame one, into the sea, to give them, as was said, a chance for their lives. Three years afterwards, this same turtle, with its three fins, and the marks of the hot iron beneath, was found again upon Ascension Island.

It is sufficient to refer to the results of numerous experiments which have been instituted on the fry of the Salmonidae, to be convinced of the prevalence of the same surprising impulse also among fishes.

In the invertebrate animals, we have, apparently, proof of the existence of this principle, in the fact of the great distances to which many hymenopterous insects are known to range for food. A decisive experiment, however, is still needed to render the inference conclusive; and I venture to suggest, to whoever may have the opportunity and inclination, that of marking a number of bees from the same hive, and suffering them to fly from, say, a hundred miles' distance. There is hardly a doubt that they would be found to regain their abode; and it would be interesting to ascertain the time they would require to do so.

Some migratory birds are observed to resort annually to the exact same winter quarters; for illustrative proof of which, refer to Bewick's description of the woodcock. Other species would seem to wander through the winter, of which the waxwing may be cited as an example. They all, however, appear to return to their former breeding haunts, where dispersion is effected, in those species which do not nidificate in society, by the older individuals (which are always the first to return) driving away their young of a former year; which latter, however, do not commonly retire farther than they can help, as I have had occasion to notice in some instances.

The bearings of this law on the geographical distribution of species do not appear to have been sufficiently taken into consideration. For instance, Mr. Selby remarks, as an extraordinary circumstance, tending to show within what abrupt boundaries the natural range of particular species is confined, the abundance of the white stork in Holland, and its excessive rarity on the opposite English coast. In Holland, be it remembered, it meets with encouragement; whereas, in this country, no sooner does an individual make its appearance, than it is immediately shot down. Once allowed to settle, it would doubtless soon colonize our fenny counties.

Some years ago, a considerable flock of spoonbills settled in a part of Aberdeenshire; whereupon the whole neighborhood uprose in arms, till every bird of them was killed. Here, probably, we have an instance of another phenomenon in the animal world, which should not be overlooked in this treatise. When a species increases numerically in any habitat beyond what the latter is adequate to sustain (a circumstance which, in the higher groups, can hardly happen, except in those of social habits), either their ranks are mysteriously thinned by what is termed *epizooty,* or an erratic impulse (unrestrained by the localizing principle we have been considering) instinctively prompts a portion of them to seek fresh quarters. This is observed more in mammalians than in birds, but is particularly noticeable in the insect tribes; various species of which, though solitary in their usual habit, have been known to assemble at times in prodigious multitudes, prompted by a general impulse, which, however, appears to be less conferred with intent to extend the previous range of their distribution, than to preserve the species within due bounds in its native locality; for the numerous dangers with which these wanderers are necessarily everywhere beset absolutely appear to suffice, in most instances, to prevent their permanently establishing themselves in other places; a remarkable fact, notorious to all who have attended to the subject. So many causes are there in operation which combine to circumscribe the geographic range of species.

A variety of important considerations here crowd upon the mind; foremost of which is the inquiry, that, as man, by removing species from their appropriate haunts, superinduces changes on their physical constitution and adaptations, to what extent may not the same take place in wild nature, so that, in a few generations, distinctive characters may be acquired, such as are recognized as indicative of specific diversity? It is a positive fact, for example, that the nestling plumage of larks, hatched in a red gravelly locality, is of a paler and more rufous tint than in those bred upon a dark soil.[17] May not, then, a large proportion of what are considered species have descended from a common parentage?

I would briefly despatch this interrogatory, as able writers have often taken the subject in hand. It is, moreover, foreign to the professed object of this paper. There are many phenomena which tend, in no small degree, to favor the supposition, and none more so than what I have termed the localizing principle, which must occasion, to a great extent, what is called "breeding in and in," and, therefore, the transmission of individual peculiarities. We have seen, however, the extreme difficulties which most species have to encounter when occurring beyond the sphere of their adaptations; difficulties which must require human aid, in general, to render surmountable. But, without re-entering into the details of this subject, it will be sufficiently clear to all who consider the matter, that, were this self-adapting system to prevail to any extent, we should in vain seek for those constant and invariable distinctions which are found to obtain. Instead of a species becoming gradually less numerous where its haunts grade imperceptibly away, we should discover a corresponding gradation in its adaptations; and, as the most dissimilar varieties of one species (those of the dog, for instance) propagate as readily together as individuals of the same variety, producing offspring of blended characters,[18] so much so, that human interference is requisite to preserve a breed unadulterated, the unbending permanency of the distinguishing characteristics of all wild animals becomes of double import. Moreover, the characters in which these differ are of a diverse kind from these observable in any but the most distant of mere varieties; for they rarely agree in the relative proportions of parts, which are the most fixed of all specific distinctions. It is, therefore, advisedly that we are enabled to state that the raven of the Cape is distinct from the raven of South America; that both are again different from that of the South Sea Islands and from that of Europe. The common jay is diffused over a wide range of latitude. but is the same in Italy as in Sweden: this would not be were it affected by locality or climate; the very trivial distinctions, therefore, which characterize it apart from that of Japan, and from that of Asia Minor, we are warranted in esteeming of specific value. Until the jays of intervening localities present inosculant characters or until precisely analogous diversities are, in wild nature, observed to be produced by locality or climate, the above conclusion is as irresistible as it is incontrovertible.[19]

When, too, we perceive that species so very general in their adaptations as the typical Corvidae are limited in their range, it behooves us to be most cautious in assuming the specifical identity of the most similar animals from widely separated localities. Let it be remembered that no reason can be assigned why those originally distinct should not exactly resemble. Human agency apart, I do not think there is a single species which even approximates to universal distribution. Of course, we can only judge from probability and analogy. Great locomotive power, even the maximum, by no means necessarily implies an extensive distribution: witness the common swift, and its American analogue *(Chaetura pellasgica),* neither of which have been known to straggle across the Atlantic, like many birds of far less power of wing, but are bound by the localizing principle. It is true, this principle can apply only to such species as are locomotive; but it is equally true, that other causes analogously restrain the undue diffusion of those which are individually fixed. Thus we hear of the agency of sea currents in transporting seeds, which must abundantly be carried out into the ocean by the action of rivers; but it appears not to have been remembered that steeping in sea water destroys the vital principle; that moisture induces germination, which, once excited, can only be checked by the final cessation of the vital functions.[20] Analogy would lead us to infer that such antagonist principles obtain throughout creation, whether or not human observation may have yet detected their existence. It would be easy to point out additional hindrances to the more extensive spread of species of fixed habit, by treating on the fraction which are allowed to attain maturity, even in their normal habitat, of the multitude of germs which are annually produced; and in what ratio the causes which prevent the numerical increase of a species in its indigenous locality would act where its adaptations are not in strict accordance will sufficiently appear, on considering the exquisite perfection of those of the races with which it would have to contend. If there is a probability that any species has become naturally of general distribution, it is in the case of two lepidopterous insects, *Acherontia atropos* and *Cynthia cardui,* both of which are of peculiarly erratic habits; and it is said that these are found throughout the world: yet the differences which exist in specimens from diverse localities are hard to reconcile with specific variation, at least to judge from what tropical specimens I have seen of the former; and an eminent entomological friend has remarked to me, in conversation, that he is equally sceptical, judging from his own experience, of many *Cynthiae* assumed to be *cardui*. It will be borne in mind, however, that man has unintentionally carried with him the seeds of the very prolific plants on which the painted-lady butterfly feeds, wherever he has introduced the Cerealia.

But to return to that mysterious guiding principle, so important, as we have seen, in regulating the distribution of species; and which I have asserted to be not wholly absent from the human constitution. It has been stated of many savages, and more particularly of the aborigines of Australia, that they are enabled to return for even hundreds of miles to their homes, though totally unacquainted with the route, being led by an intuitive impulse that they cannot explain. This seems incredible: but we know that diurnal birds will return by night from the heart of Africa to their former abode, marked individuals having done so; and we also know that a pigeon, carried from Paris to Constantinople, has flown back to the former city: these facts will tend to diminish our scepticism. I have two instances, however, of the manifestation of this principle by Europeans, when in a state of insensibility; for both of which I am indebted to the parties themselves, gentlemen of unimpeachable veracity: both of them returned, in this condition, to their temporary homes (one in the dark, and for upwards of a mile, having been thrown from his horse, which remained on the spot till found next morning), by routes with which they were quite unacquainted. I am not disposed to enlarge at present on this subject, by inquiring to what extent numerous phenomena recorded of somnambulists may be explicable on this obscure principle. We hear continually of surprising instances of blind men finding their way, with a degree of accuracy very difficult to comprehend; and, also, of drunkards stumbling home, when apparently unobservant of external objects. It will be sufficient if these hints serve to awaken the reader's attention, and so, peradventure, elicit some additional facts.

We have now traced to their ultimate results certain of the bearings of the intuitive information conferred on brutes, which, in wild nature, mainly influences their actions. We have seen that man is denied innate knowledge of the properties of objects, and is, therefore, necessitated to observe and reflect; in a word, to learn. Hence the necessity of a long infancy and superior intelligence; hence that progressiveness which so eminently distinguishes him from all other races. I have nowhere denied that other animals are capable of reflection; but I assert that, unrestrained by human influence, their inherent instincts sufficing to insure their weal and maintenance, these, in consequence, supersede the necessity of habitual observation; whence their reasoning even may be independent of experience. Indeed, it is hard to instance a case wherein the conduct of truly wild animals may not be satisfactorily referred to instinctive motives; but that such cases do occur is shown by eaves swallows *(Hirundo urbica)* having been known to immure a sitting sparrow that had usurped their nest; which fact is proved sufficiently to be in nowise referable to instinct, inasmuch as it is contrary to the ordinary habit of the species upon such occasions.[21] It will be readily admitted, however, that such instances are extremely rare exceptions to the general rule; and I imagine there are few who will be disposed to refer the ordinary habits of any species of the lower animals to aught else than original intuition.

I have yet another phenomenon, which is now, I believe, for the first time introduced to notice. It is the occurrence, in domesticated animals, of what is analogous to idiocy in the human race. Of this I have several instances in poultry, and one in a sheep. It consists in the privation of more or less of that intuitive knowledge which is needed to enable an animal to maintain its existence amid the numerous dangers with which it is naturally surrounded; dangers against which no experience could suffice to fortify it. The creatures I allude to evinced a listlessness in their deportment strikingly similar to what is commonly observed in human idiots: they sought not the society of their companions; and one of them, a hen (of which only I can speak from personal observation), would heedlessly wander close before the kennel of a fierce dog, which the other fowls constantly avoided. Whether the dog would have attacked another fowl, I cannot say; very likely not: but it is a well known fact, that the most savage of the canine race will never attack a human idiot, nor a child, nor a person stupefied by intoxication: of the truth of which latter, a most remarkable exemplification lately happened in this neighborhood; a drunken stranger having been absolutely permitted to share the straw of a very fierce watch-dog, which those it knows can hardly approach with safety.

In the foregoing pages, I have all along been considering the diversity of human influence from that of all other organized races, rather in its effects towards these latter, than by taking the higher ground of natural theology in reference to human kind, and recognizing, in the grand aggregate of all that has been effected in past ages by the joint influence of every cause that has been in operation, not only a gradual prospective adaptation to the welfare of each succeeding race, but an ulterior object in capacitating the globe for the residence of human beings. A new era commenced with the introduction of man upon this world: a secondary intelligence was permitted to assume the dominion over matter, in so far as, by experimenting upon its properties, it can elucidate the unvarying laws which regulate these, a knowledge of which is indispensable to empower intellect to direct their operation.[22] To man it was given to "conquer the whole earth and subdue it"; and who can venture to aver the ultimate limits of those changes which he everywhere superinduces; changes which, in conjunction with the physical laws which wear away the land and uplift the bed of the ocean, may, in time, be gradually fatal to the normal condition of every other race, and to the existence even of by far the greater number? that is, assuming, what there is every reason to infer, that the human species was the last act of creation upon this world, and that it will continue to be so until its removal. It is needless to add, that a prodigious lapse of time is here required; and, to judge from data which the past history of the globe abundantly furnishes, in legible records, wherever we turn our eyes; to judge from the progressiveness of human intellect, and the long, long while that must yet transpire ere man can hope to assume that rank, as a consistent being, for which his faculties clearly show that he was intended, the duration of his existence upon this planet would appear likely to bear proportion to that immense period that the globe will continue fitted for his reception; a period, it may be presumed, that will abundantly suffice to alternate the land and sea, as we know has repeatedly happened heretofore, and which may sweep from existence the inhabitants of the present ocean, as those of which the exuviae occur in the chalk have become extinct before them.[23] The past affording the only record from which we are competent to judge rationally of the future, this inverse analogy would argue a continuance of the refrigeration of our planet, till it shall be again unfitted for the existence of organic beings. As by a catenation of obvious and palpable facts we can trace back the history of this world to a state of fusion, if not of general nebulosity, so are we warranted in anticipating its future congelation, when old age shall have come over it as barrenness, and the gases shall have solidified by intensity of cold; when, it may be, the sun himself shall have grown dim, and nature, in so far as this system is concerned, have sunk in years; when the stupendous cycle shall have been accomplished. Then, and with humble reverence let the mighty acts of Supreme Omnipotence be spoken of, it may be that the eternal and ever-glorious Being which willed matter into existence shall pronounce on it the final doom of annihilation; and

the great globe itself
. . . shall dissolve,
And like this insubstantial pageant faded
Leave not a rack behind . . .

Or, what is far more probable, to judge from the universal analogies of all that is within our grasp, its elements shall again be called forth into light and life, and blaze forth the recommencement of the same system.

It is inconsistent with our notions of divine benevolence, to suppose that the human race shall be suffered to linger here upon earth, till such secondary causes as we have been considering should suffice to gather the last man to his fathers.

PART

The Forgotten Parent

Blyth died in London on December 27, 1873. Several years later, a biographical sketch by his long-time friend Arthur Grote was published in the Journal of the Asiatic Society of Bengal. *In it Grote drew on Blyth's letters, which are now missing, as well as on his own recollections. Other than Grote's memoir, which follows, the obituaries of Blyth are of little value.*

Edward Blyth

MORE THAN ONE obituary notice of Blyth and his scientific labors, by competent and kindly pens, has already appeared in the columns of those journals to which he had been in the habit of contributing. This seems a fitting place for collecting in a brief memoir such particulars as are obtainable of his early life, and such as I can myself supply of his long career in our Society's service. My personal acquaintance with him commenced on my return to India from furlough in 1848. He had then been curator of our museum for seven years, and my official connection with the Society, combined with a taste for his pursuits, brought me into frequent and close relations with him. Of the incidents of his pre-Indian life some knowledge has been obtained from his sister, who has kindly given me access to such of his letters as are in her possession.

Edward Blyth was born in London on the 23rd December, 1810. His father was of a Norfolk family, and from him the son appears to have inherited both his taste for nature and the retentive memory for which he was so remarkable. Blyth's father died in 1820, leaving four children, whose care and education now devolved on the widow, a Hampshire lady, who at once sent Edward, the eldest boy, to Dr. Fennell's school at Wimbledon.

Here the boy seems to have made unusual progress in his books, but the school reports describe him as of truant habits, and as being frequently found in the woods. He left school in 1825, and his mother seems at first to have intended him for a university career, and ultimately for the Church, but at Dr. Fennell's suggestion she sent her son to London to study chemistry under Mr. Keating, of St. Paul's Churchyard. He did not, however, long persevere in this study, being dissatisfied with his instructor's mode of teaching. His enthusiasm for natural history pursuits disinclined him for any ordinary employment, and on coming of age he embarked the little means he had in a druggist's business at Tooting. To this he seems to have given little personal attention. The management of the business was left to another, while Blyth devoted all his time to the study which engrossed his thoughts. "Never," says his sister, "was any youth more industrious; up at three or four in the morning, reading, making notes, sketching bones, coloring maps, stuffing birds by the hundred, collecting butterflies, and beetles—teaching himself German sufficiently to translate it readily, singing always merrily at intervals." He took a room in Pall Mall, to have readier access to books, and passed much of his time in the British Museum, in which, or in some kindred institution, he tried hard to find employment.

Naturally the Tooting business did not thrive under such fitful management. Blyth soon found himself in serious difficulties; such literary work as offered itself in his own special line of study supplied him with but precarious means. In the introduction to his edition of White's *Selborne*, which bears date from Lower Tooting, 1836, he alludes to the anxieties which then surrounded him, though "his mind," he adds, "cleaves to its favorite pursuit in defiance of many obstacles and interruptions, and eagerly avails itself of every occasion to contribute a mite to the stock of general information." Young as he was, Blyth had at this time earned for himself a reputation as a diligent and accurate field observer, and he corresponded with many of the leading naturalists of the day. He seems to have been a contributor to both Loudon's and Charlesworth's series of *The Magazine of Natural History* from 1833 till his departure for India, and in one of his papers of the volume for 1838 he proposed a new arrangement of Insessorial birds. Rennie enlisted him as a writer in the *Field Naturalist,* and he was associated with Mudie, Johnston and Westwood, in an illustrated translation of Cuvier, which was published by Orr and Co. in 1840. Blyth undertook the mammals, birds, and reptiles in this work, adding much original matter of his own, which is inclosed within brackets. A new and enlarged edition of the work appeared in 1854, with important additions to the molluscs and fishes by Dr. Carpenter.

The *Proceedings of the Zoological Society* from 1837 to 1840 contain a few papers read by Blyth at their meetings. One of these, on the osteology of the Great Auk, observes on the distinctive characters of auks and penguins. In another he draws attention to peculiarities in the structure of the feet of the trogons. But the most important of these contributions was his monograph of the genus *Ovis,* read in 1840.[1] He here describes fifteen species of sheep, including the then newly discovered *O. poli,* from Pamir. At the same meeting he exhibited drawings and specimens of the yak, Kashmir stag, markhur, Himalayan ibex, and other Indian ruminants, his remarks on which show the attention which he had already begun to give to the zoology of India.

Just at this time our Society had obtained from the Court of Directors a grant for a paid curator of its museum, which had grown into a collection beyond what was manageable by the honorary office-bearers who had hitherto looked after it. The labors of Hodgson, Cantor, M'Clelland and others, had filled it with valuable zoological specimens, which with important fossil and other contributions were falling into great disorder. Professor H. H. Wilson, then our honorary agent in London, was asked to select a competent man to undertake the general charge of the museum, and the appointment was offered to and accepted by Blyth, then in weak health, and professionally advised to seek a warmer climate. Provided with passage and outfit by the Court of Directors, the latter arrived in Calcutta in September 1841. His letter to Mr. H. Torrens, published in our Society's *Proceedings* for that month (vide *Journ.* 10:2:756), expresses the diffidence with which he entered on the charge of the mineral department of the museum; but of this duty he was largely relieved in the following year on the appointment of Mr. Piddington to all the departments of economic geology. He still retained the custody of the paleontological specimens.

One of the duties impressed on him by our then president, Sir E. Ryan, was that of furnishing monthly reports at the Society's meetings; and in October 1841, he accordingly submitted the first of that long series of useful reports which appear in our *Proceedings* with scarcely any intermission for the next twenty years. Each of the monthly issues of this journal for the remainder of 1841 contains a paper by Blyth. In the first of these, "A general review of the species of True Stag," etc., he committed himself to an opinion, shared with him by Ogilby, regarding Hodgson's *Cervus affinis,* which, as Jerdon has pointed out (*Mamm.* p. 252), he did not recant till 1861.

Many of Blyth's reports fill from fifteen to twenty pages, and his remarks on the various contributions which reached him were just what were wanted by the field observers who supplied them. The active correspondence which he set on foot with these and with sportsmen, all more or less naturalists, throughout India, encouraged their useful pursuits, and brought him a large accession of specimens. He received in July 1846 the thanks of the monthly meeting of our Society for his exertions "in opening out new channels of scientific intercourse."[2] He had already found it necessary to apply for assistance in his museum duties, but the Society had not the means of supplementing the Government grant beyond the small allowance which they gave him for house rent. Had Blyth been less devoted to the special service in which he had engaged, there were not wanting to him opportunities of finding far more remunerative employment in other quarters. The Dutch authorities in Java seem to have about this time made him a very tempting offer.

The *Proceedings of the Zoological Society* for 1841 and 1842 contain two letters from Blyth, of which one was written on the voyage out to India, and the other shortly after his arrival.[3] The latter contained remarks on various species of birds found in India and Europe. Nothing from his pen appears in the *Calcutta Journal of Natural History,* of which the publication had just commenced when he reached India, and which was brought to a close in 1847. He found time, however, to send home several papers for the *Annals of Natural History* in 1844–48.

The unpleasant episode in regard to the publication of the Burnes zoological drawings with Dr. Lord's notes had occurred before I joined the Society. The materials, which consisted of certain wretched figures by a native artist, and some descriptions of already well-known species, the Afghanistan localities of which were alone new, had been made over to us by the Government before Blyth became our curator. The lithographer's death had brought the work to a stand, and when inquiry was made in 1844, the notes which were to furnish the letter-press were not forthcoming. Blyth's explanation of his share in their disappearance will be found in our *Proceedings* of October 1844.[4] This was followed by a controversy with Mr. Torrens, then our secretary; and the financial embarrassments of the Society soon afterwards necessitated the abandonment of the publication.[5] Of the fourteen colored copies of the completed plates, I possess one, and I quite agree with Blyth that their issue would have brought ridicule on the Society.

The heavy outlay incurred on this undertaking, and on the publication of Cantor's Chusan drawings, was unfortunately the cause, not only of the embarrassments just noticed, but of a temporary estrangement between the philological and physical classes of our members. Funds which had been assigned by the Government for furthering Oriental literature had no doubt been appropriated to other objects. Blyth came in for a share of this discontent on the part of the Orientalists, and some naturalists also complained that he was enriching the mammal and bird departments of the museum at the expense of those of the shells, fossils and insects. The want, too, of a catalogue of the collections had been long felt, and the curator had been repeatedly urged to supply it. The Council refers to his delay in performing this duty in their report of 1848, while commending "his regularity of attendance and remarkable industry."[6] His application for increased pay and a retiring pension was referred to the Society at large with the following guarded remarks:

> It must be admitted that for any scientific man capable of discharging the duties on which Mr. Blyth is employed, and of performing them with activity and zeal, for the advancement of science, etc., the [monthly] salary of 250 rupees is a very inadequate compensation. But the Council cannot but regard the present as an inauspicious period to address the Honorable Court in furtherance of any pecuniary claim. The diversion of the Oriental grant to so large an amount as has but lately been brought to notice, cannot be regarded with indifference by them, nor can it have disposed them to entertain with much favor any fresh demand on their munificence preferred by the Society.

The application was then referred for report to the natural history section, and notwithstanding the stout struggle made on his behalf in the section, their report was unfavorable to Blyth's claims, which were finally negatived at the July meeting in 1848.[7]

In the following year Blyth published his *Catalogue of Birds,* which had in fact long been ready for issue in a form which would have satisfied the Council. It had been constantly kept back for the appendices, addenda, and "further addenda," which disfigure the volume, and seriously detract from its value as a work of reference. This habitual reluctance of his to part with his compositions till he had embodied in them his latest gained information is conspicuous throughout his contributions, and it is in fact partly due to this habit that these Burman catalogues form a posthumous publication.

Blyth availed himself of every opportunity which offered of escape from his closet studies to resume his early habits of field observation. Frequent mention will be found in his reports of the little excursions into the country which he thus made, and of the practical results obtained from them. The geniality of his disposition and the large store of general information at his command insured him a warm welcome in all quarters. One of his favorite resorts was Khulna, on the edge of the Jessore Sunderbuns, where the indigo factory of an intelligent and untiring observer[8] offered him a favorable station for field pursuits.

Several contributions from Blyth on his special subject will be found in the pages of the different sporting journals which have appeared in Calcutta. He was on the regular staff of the *Indian Field.* In the *India Sporting Review* he published a sketch of "The Osteology of the Elephant," and a series of papers on "The Feline Animals of India." For the *Calcutta Review* he wrote an article on the "Birds of India." It gives the results of his latest experience on the subject of the communication made in 1842 to the Zoological Society, which has been noticed above, and shows that of 353 species of birds admitted by Yarrell into the English avifauna, no less than 140 are found in India.

In 1854 Blyth was married to Mrs. Hodges, a young widow whom he had known as Miss Sutton, and who had lately come out to join some relatives in India. This step on his part necessarily aggravated the embarrassments entailed on him by his inadequate income, and on completing his fourteenth year of service in 1855, he memorialized the Court of Directors for an increased salary and for a pension "after a certain number of years' service." In the second paragraph of his memorial he observes, "that however desirous the Asiatic Society might be of augmenting your memorialist's personal allowances, the ever-increasing demands on its income, consequent on the extension of its collections among other causes, altogether disables it from so doing." On this memorial being submitted to the meeting of May 1856, it was agreed to forward the document to Government, "with the expression of the high sense entertained by the Society of the value of Mr. Blyth's labors in the department of natural history, and of its hope that the memorial may be favorably considered by the Honorable Court."[9]

The extract just given will show, in Blyth's own words, that he had no complaints to make of our Society's treatment of him. Mr. A. Hume, who seems to have first joined our Society in 1870, has gone somewhat out of his way in his "Rough Notes" to do justice to Blyth's merits as curator, at the expense of older members.[10] The language used is in Mr. Hume's characteristic style, and is as offensive as the charge brought against the Society is unjust. The same charge is implied in the use of the words "neglect and harshness" in the "In Memoriam" with which Vol. 2 of *Stray Feathers* opens, and which, with this exception, describes with much truth and feeling the life-long struggle in India, as at home, which Blyth's scientific ardor supported him in maintaining against the most depressing obstacles.

That nothing came of this memorial is due probably in some measure to the movement which commenced in 1857 for transferring our collections to an imperial museum, but mainly to the great convulsion which shook our empire in that year. I find no record in our *Proceedings* of any reply having been made to our recommendation, and the negotiations for the foundation of the new museum were not resumed for some three years.

Blyth made a short tour in the Northwest Provinces in July 1856. He spent some six weeks in Lucknow, Cawnpore, Allahabad and Benares. Oude had just been annexed, and the sale of the Royal Menagerie at Lucknow had been determined on. The tigers were the finest caged specimens in the world, and to one who understood their value in the European market, the inducement to buy and ship the animals was irresistible. A German friend joined in the speculation, and found the necessary funds. Blyth was to do the rest, and as no competitors offered, he bought the bulk of the collection for a trifle. Eighteen magnificent tigers were sold at 20 rupees (£2) a head! Some casualties occurred on the passage down the river; but his collection, when exhibited in Calcutta, contained sixteen tigers, one leopard, one bear, two cheetas, three caracals, two rhinoceroses, and a giraffe, which carried a saddle and was daily ridden. Difficulties unfortunately occurred in finding ships for the transport of the animals, and their detention in Calcutta caused further casualties and heavy charges, which his partner would not face. The speculation collapsed, but one of the tigers which reached England realized £140.

In December 1857, Blyth had the misfortune to lose his wife. His short married life had been of the happiest, and the blow fell heavily on him. His letters to his sister for the early months of 1858 are painful to read. The shock proved too much for him, and brought on a serious attack of illness; it threatened paralysis of the heart, and he seems to have been subject to partial returns of similar attacks for the rest of his life. His health too suffered much from the isolation imposed on him by his straitened means, and from want of proper exercise. Some distraction for his thoughts was luckily afforded at this time by the opening up of a new fauna in the Andaman Islands, which Dr. Mouatt had been sent to report on before their occupation as a penal settlement. To this Report Blyth contributed an interesting chapter on the zoology of the Islands, so far as it was then known.

The China expedition of 1860 was considered both at home and in India a good opportunity for obtaining information regarding the natural history of North China. Blyth's name was put forward as that of a naturalist readily available and eminently qualified for the post of naturalist to the expedition. Replying to Lord Canning's objections that scientific observations in a hostile country would have to be carried on at much personal risk, our Council,[11] while urging the importance of the mission in a scientific point of view, stated on Blyth's behalf that "he was quite willing to encounter the danger, whatever it might be." The application, however, failed: no naturalist was appointed. This result was to be regretted, as it affected Blyth personally, for his health was failing, and the sea voyage, with the stimulus afforded by so interesting a mission, would have been most beneficial to him, and would probably have averted the utter breakdown which was now at hand. It is doubtful whether he was equal to the more laborious task which he offered to undertake in the following year, when the scientific expedition into Chinese Tartary was projected by the Government.

Blyth was a staunch adherent of Darwin's views, and an opportunity of thus declaring himself offered at our November meeting in 1860, when Mr. H. Blanford read his paper on the well-known work of Dr. Broun on the laws of development of organized beings. The value attached by Darwin to Blyth's observations is shown by the frequent reference made to them, more especially in his *Animals and Plants under Domestication.* His first citation of Blyth in the latter work describes him as an "excellent authority," and the many quotations that follow in these interesting volumes show how carefully he read and noted all that fell from Blyth, even in his contributions to sporting journals.

In 1861 Blyth's health fairly gave way, and in July of that year a second memorial was submitted to Government with a view to obtaining a reconsideration by the Secretary of State for India of his claims to a pension.[12] Lord Elgin, the New Viceroy, took up the subject warmly, and pressed it on the attention of the Home authorities as a special case: "the case," as he observed, "of a man of science who had devoted himself for a very small salary to duties in connection with the Asiatic Society, a body aided by and closely identified with the Government of India, from which the public have derived great advantage."[13] After describing Blyth as "the creator of the Natural History Museum, which has hitherto supplied the place of a public museum in the Metropolis of India, and which will probably soon be made over to Government as part of a national museum," and referring to the importance of Blyth's labors in zoology in maintaining and extending the character and standing of our Society, this dispatch concludes thus: "His Excellency in Council considers, therefore, that if under such circumstances Mr. Blyth should, after twenty years' service, be compelled to retire from ill health, brought on very much by his exertions in pursuit of science, it would not be creditable to the Government that he should be allowed to leave without any retiring pension."

Meanwhile, Blyth was only enabled to remain at his post by the facilities which the Council afforded him of making short successive visits to Burma. He was for some five months in that province, from which, and more especially from the Yonzalin River, he communicated several interesting letters. His camp life there agreed with him, and he had kind friends like Phayre, Fytche and Tickell to associate with and take care of him. His return to Calcutta was always attended by a relapse, and the hot season of 1862 brough him to a state for which there was no alternative but instant departure for Europe. As yet, however, no orders had been received from home in regard to the pension. It was clear that for these it would not do to wait, and the Council under the emergency gave Blyth a year's leave on full pay.[14] He had hardly gone when the expected reply was received and this, notwithstanding the Viceroy's strongly expressed opinion, proved an unfavorable one. Eventually a pension of £150 a year was conceded, owing, I believe, mainly to the untiring efforts made in London on Blyth's behalf by the late Sir P. Cautley and Dr. Falconer.[15]

By the end of 1864 our Society's negotiations with the Government for the transfer of its collections to the Indian Museum had been brought to a successful close, and at the November meeting the following just tribute was paid to our late curator in the form of a resolution, which, on the Council's proposition, was carried unanimously:

> On the eve of transferring the zoological collections of the Society to Government, to form the nucleus of an imperial museum of natural history, the Society wishes to record its sense of the important services rendered by its late curator, Mr. Blyth, in the formation of those collections. In the period of twenty-two years during which Mr. Blyth was curator of the Society's museum, he has formed a large and valuable series of specimens richly illustrative of the ornithology of India and the Burmese peninsula, and has added largely to the mammalian and other vertebrate collections of the museum; while, by his numerous descriptive papers and catalogues of the museum specimens, he has made the materials thus amassed by him subservient to zoological science at large, and especially valuable to those engaged in the study of the vertebrate fauna of India and its adjoining countries.[16]

Blyth was elected an honorary member of the Society in the following year. The museum was now under a Board of Trustees, and a new curator, better paid, and with all the prospective advantages of a government official, had taken charge of it. Writing to me from Malvern, in June 1865, Blyth says: "I had always a presentiment that my successor in the museum would be more adequately remunerated, beginning with just double what I had after more than twenty years' work, with an additional £50 yearly, and house accommodation! How very much more could I have accomplished with such an income!" With this mild explosion he brushed off discontent, and strove to make the most of his small means. His letters to me, and these were frequent up to the time of my leaving India in 1868, were full of his own special subject; some of them are published in our Society's *Proceedings.*

In January 1864, Blyth visited Dublin, where he read two papers before the Royal Irish Academy. The first of these was "On the True Stags or Elaphine division of genus *Cervus,*" and does not appear to have been printed *in extenso* in the Academy's *Proceedings* (8:458, Jan. 11, 1864). His other paper, "On the Animal Inhabitants of Ancient Ireland," was published at length in the Academy's *Proceedings* (Jan. 25, p. 472). What the extraordinary bones were which he exhibited at the meeting, and which he referred to as "probably Tibetan," was not explained in any of his letters.

At a meeting of the Geological Society of Dublin, he made some remarks on a paper of Professor Haughton's "On Geological Epochs," and expressed his concurrence in Dr. Carte's identification of the bones of the polar bear discovered in Lough Gur, in County Limerick.[17] On further examination, however, these bones have been pronounced by Mr. Busk to be indistinguishable from those of *Ursus ferox.*

The question of zoological distribution will be found to have been treated by Blyth, in a paper which he contributed to *Nature* in 1871 (March 30). He had been led to consider it while drawing up the introductory chapter which was to preface these catalogues, for in a letter to me dated 15 July of that year he refers to this MS. as follows:

> I suppose that Phayre showed you my sketch of what I conceive to be the true regions and sub-regions of S. E. Asia, and I expected that he would have modified somewhat my notions with regard to the provinces into which I venture to divide the *Indo-Chinese sub-region,* but he seems to have assented to them altogether. Only yesterday I received the *Proceedings of the Asiatic Society* for April and May last, and the *Journal of the Asiatic Society* for April and May last, and the *Journal of the Asiatic Society of Bengal,* Part 2, No. 1, 1871, and in p. 84 of the *Proceedings* I find some remarks by Stoliczka which quite confirm my views, only that I think that, with regard to the extension of the Malayan

> fauna into India, he should rather have said *Southern* India, because the African affinities of Central and Northern India, inclusive of the Siwâlik Deposits, are of ancient date, as shown by the occurrence of *Bos namadicus* in Central India, which is barely separable from the European *B. primigenius* (a type of *Bos* which is elsewhere only known from Europe), and by the presence of giraffes and of antelopes of African type in the Siwâlik Deposits. I have such an enormous mass of valuable facts to deal with, that I gave over making them public in driblets at the meetings of the Zoological Society; and I have now time and undisturbed leisure to treat of them in a work which I am preparing on "The Origination of Species," a subject upon which I think I can throw some light.[18]

As pointed out in a note, Blyth's "Austral-Asian region" is generally the same with Dr. Sclater's "Indian region," minus Hindustan proper or the plains of Upper India east and south of the northwest desert—the Dukhun or tableland of the Peninsula with the intervening territory, inclusive of the Vindhyan Ghats—the Coromandel Coast and the low northern half of Ceylon—all of which Blyth places in his Ethiopian region. What remains of India after this large deduction Blyth distributes through three sub-regions, viz., the Himalayan, Indo-Chinese, and Cinghalese. India cannot, he argues, be treated as a natural zoological province: it is a borderland in which different zoological regions meet, and one, therefore, "of extraordinarily complex zoological affinities." Burma of course falls within his Indo-Chinese sub-region, which extends southward as far as Penang and Province Wellesley, where his Malayan sub-region commences.

The interest which Blyth had always taken in the rhinoceros group was revived by the safe arrival at the Zoological Gardens of the Chittagong individual, the *Ceratorhinus crossei* of the present catalogue. In his paper contributed to the *Annals* in 1872, he argues against Gray's assignment of this species to *Rhinoceros sumatrensis,* and in favor of its identity with the fine Tavoy specimen shot by Col. Fytche, and figured in this *Journal,* Vol. 31, p. 156. Blyth's conjecture that the Arakan Hills is one of the habitats of this species is borne out by the letter in which Capt. Lewin, the superintendent of the Hill Tracts of Chittagong, first reported to me in 1867 the capture of the animal.[19] After giving her measurements, which were then six feet from crown of head to root of tail, and four feet two inches in height, and otherwise minutely describing her horns, Capt. Lewin adds: "You are mistaken I think in supposing that she has come from the Tenasserim Provinces—the two-horned species is found in my hills. I have seen one alive, and several of my men have seen a dead one."

In the *Journal of Travel and Natural History,* No. 2 (p. 130), of 1868, will be found a letter from Blyth in explanation of some remarks which he had made at the Zoological Society on the occasional shedding or loss by violence of rhinoceros' horns, followed by their renewal. In this he takes the opportunity of pointing out the tendency which some species have to develop a rudimentary horn on the forehead, and argues for the possible explanation in this manner of cases of three-horned rhinoceroses being reported by travelers.

The connection which Blyth established, first with *Land and Water,* and later with the *Field,* gave him interesting literary occupation; and the "Naturalist" columns of both these journals abound in scraps by "Zoophilus," which did real service to the advancement of scientific truth. No pen so ready as his to expose current fallacies or sensational announcements in works of travel of the results of loose and careless observations. Very many of his "scraps" are worthy of being collected and preserved, for such use as we see they have been turned to by Mr. Darwin. These columns occasionally contained more elaborate papers, such as the series in the *Field* for 1873, on "Wild Animals Dispersed by Human Agency," and "On the Gruidae or Crane Family." This monograph, for such it amounts to, was its writer's last utterance. He had long been ailing, and in the autumn of this year he became very ill, and went to Antwerp for a change. On his return he called on me, feeling, as he said, better, though complaining of great prostration. He seemed full of what he had seen in the Antwerp Zoological Garden, where he thought he had found another new species of Rhinoceros. This was our last interview. Though nursed by a tenderly attached sister, his weakness increased, and he died of heart disease on the 27th of December, within a day or two of his sixty-third birthday.

More competent authorities than I can pretend to be have done justice to the high intellectual powers which Blyth displayed from the outset of his career as a naturalist, to the wonderful capacity and accuracy of his memory, which, unassisted by any systematic notes, assimilated the facts once stored in it, and enabled him readily to refer to his authority for them; to his great power of generalization, and to the conscientious use which he made of it. Abundant proof of the high respect with which his opinions were always listened to, and of the careful consideration given to them even where they were not accepted, is to be found in the published works of his brother naturalists. No higher testimony to his habitual scientific caution need be adduced than that of Mr. Darwin, but it is equally borne by Jerdon throughout his published writings. Gould refers to him as "one of the first zoologists of his time, and the founder of the study of that science in India."[20] I confine myself here to putting on record the tribute of an old and intimate friend, to the excellent qualities of heart possessed by Blyth. The warmth and freshness of his feelings which first inspired him with the love of nature clung to him through his chequered life, and kept him on good terms with the world, which punished him, as it is wont to do, for not learning more of its wisdom. Had he been a less imaginative and a more practical man, he must have been a prosperous one. Few men who have written so much have left in their writings so little that is bitter. No man that I have ever known was so free as he was from the spirit of intolerance; and the absence of this is a marked feature in all his controversial papers. All too that he knew was at the service of everybody. No one asking him for information asked in vain. Among the many pleasurable reminiscences of my own long residence in India, few are more agreeable than those which recall his frequent Sunday visits to me.

PART

The Evolution of Man

Once man was thought to be a fallen angel. Then we found him to be a descended ape. But as more fossils have been discovered and studied, this idea has changed radically. Both man and the apes must have evolved independently from some rudimentary common ancestor, no longer living. Indeed, before human evolution can be traced with any certainty, still unknown forms of man, or pre-man, must be unearthed.

Neanderthal Man and the Dawn of Human Paleontology

IF THE RECORD of the rocks had never been, if the stones had remained closed, if the dead bones had never spoken, still man would have wondered. He would have wondered every time a black ape chattered from the trees as they do in the Celebes where, of old, simple forest people had called them ancestors of the tribe. He would have wondered when he saw the huge orangs pass in the forest, their bodies festooned with reddish hair like moss and on their faces the sad expression of a lost humanity. He would have seen, even in Europe, the mischievous fingers and half-human ways of performing monkeys. He would have felt, aloof in religious pride and the surety of revelation though he was, a vague feeling of unease. It is a troubling thing to be a man, with a very special and assured position in the cosmos, and still to feel those amused little eyes in the bush—eyes so maddeningly like our own.

Wherever man has been aware of monkeys, they have demanded explanation; without them we might regard ourselves as unique in the universe. With them appears the sole convincing evidence of man's relationship to the lowly world about him. The rumor of apes had passed far and wide even before the days of the great voyagers. Nor is it surprising to find, even in a Europe which, for over a thousand years after the Greek philosophers, eschewed the idea of organic change, that folklore and speculation arose around the living primates, so that parts of the Christian world evoked a second, a "black" creation to account for them.

In this view the devil and his demon associates decided to attempt a creation in imitation of the work of God. When the diabolical rites were completed and the devil gazed upon the being which he had created it was found not to be human, but an ape. Great though Satan's powers were, he could not implant a soul in his abortive offspring. The ape remains, therefore, in the words of H. W. Janson, a "kind of indistinct echo or reflection of man."[1]

But whether we emphasize the tale of the black creation, or turn to various other Christian or pagan legends of differing character, we can observe one fact eternally present wherever apes are known: they are recognized as in some manner related to man, or as being deviant men or transformed men. *Naturae degenerantis homo, figura diaboli, homo sylvestris*—whatever they are named, whatever the legend of their relationship to man, they travel down the ages with him in a companionship which is inseparable. It is as though man had a shadowy companion, a psychological Doppelgänger who had remained in the forest and yet lingered along its edge to mock his civilized brother.

His shape is confused and often seen unclearly. Sometimes it is the ape, sometimes it is an untamed wild man with a club, a lurker in woody dells and caves. Sometimes the creature moves across the modern stage in a traditional harlequin patchwork coat of leaves and tatters. Or he is the hermit with strange wisdom in a cave. He is the Abominable Snow Man whose footprints were pictured as late as yesterday's newspaper. Western civilization has never quite forgotten this hairy and primordial shadow. Compounded, perhaps, of man's suppressed subconscious longings, he is ape and man and dancing bear. He is all that remains of the lost world of the trees; of the time before the cities.

Something of this creature emerges in the early interpretations of Neanderthal man, and the skull in the little cave by the

Neander served as a kind of prism or lens through which passed at mid-nineteenth century a curious conglomeration of pre-Darwinian beliefs which, transmogrified and altered, entered the world of the Darwinists. Nothing, in short, so clearly reveals the transitional nature of Darwinian thinking as the treatment accorded this single human fossil. To define and evaluate the currents of thought which swirled around that heavy-browed calvarium one must examine at some little length the beliefs about man existing in the first half of the century, and then compare those beliefs with those which obtained after the publication of the *Origin of Species* and *The Descent of Man.* Strangely enough the alterations will not be as great as we commonly assume.

Space will not permit an exhaustive pursuit of each thought element to sources in some cases more remote than the eighteenth century. For this reason I shall begin by an improvisation which may enable the reader to grasp quickly a key distinction between two approaches to the evolutionary problem. The one involves the comparative morphology of the living, the other the historical morphology of the dead. Both are based upon the comparative anatomy which was brought to such heights of perfection under the guidance of Baron Georges Cuvier.

The first used of these two methods, or ladders into the past, as they might be called, is the taxonomical approach to life which in its eighteenth-century origins arose largely out of natural theology—the Scale of Nature concept which has been so extensively examined by Professor Lovejoy in his well-known book, *The Great Chain of Being.*[2]

Theologically, it was largely assumed in the eighteenth century that creation had been instantaneous, but that organisms were arranged in a continuous chain which anatomically led to man as the crowning glory of earthly life. Such views led to the search for connections, for the links between living forms. Taxonomy and comparative anatomy were, in other words, being unconsciously promoted by the religious philosophy of the times. Seeming gaps in the living scale of life were often assumed to be filled by creatures in unknown portions of the world. It was not yet clear, either geologically or paleontologically, that the world was old, or that many of its life forms had vanished long ago; the doctrine of Special Creation held the field.

If one examines the Scale of Nature concept carefully, however, it can be observed that this linked chain of development implies a position for monkeys and anthropoid apes as standing next to man on the anatomical ladder. The only thing needed to transform the Scale of Nature idea into an evolutionary scheme is to introduce within it the conception of time in vast quantities, and the further notion that one form of life can give rise to another—that the links, in other words, are not fixed but can move their positions on the scale. This theory, the dawn of a truly evolutionary philosophy, began to come in toward the close of the eighteenth century, but its supporters constituted at that time a small and largely ignored minority.

As one examines the writings of the early evolutionists, in so far as they touch upon man, it is necessary to remember that human fossils of any sort were undiscovered, and that even the possibility that forms of life *could* become totally extinct was still under debate. It is not surprising under such circumstances that the early evolutionists tended to regard the orang as a possible existing human "ancestor" or genuine *Homo alalus* who needed only speech and the refining influence of civilization to become a man.

The higher primates, actually as yet little studied but reported upon by the voyagers with much anthropomorphic embellishment, come close to complete humanization, only a grade below the Hottentot. African kings were said to have guards who resembled monkeys and, in many of the early descriptions, it is apparent that the distinction between man and ape is extremely blurred.[3] Amongst the Australians, writes one chronicler, "there was one man who, but for the gift of speech, might very well have passed for an orangoutang. He was remarkably hairy; his arms appeared of an uncommon length; in his gait he was not perfectly upright; and his whole manner seemed to have more of the brute, and less of the human species about him than any of his countrymen. The gift of speech, however, which he must, if at all, have acquired in his infancy, will not alone prevent his actually being what he might very well have passed for."[4]

The writer, in other words, implies in this last remark that speech is an invention which could be acquired by an ape. Natives strikingly different in appearance from the white man were being arranged on a kind of evolutionary scale as intermediate links between the great apes and man. Thus, though between the first decade of the nineteenth century and the discovery of Neanderthal man in 1856, there had been a vast increase in our knowledge of the fossil past, nothing had been learned of man save the likelihood of an antiquity extending into the glacial period.

The comparative anatomy of the existing primates pointed to some close relationship with man, but it was still only possible for the evolutionist to compare living man with living ape and the living races. So far as man was concerned, the ladder into the past was still the taxonomic ladder all of whose rungs were still in existence. Important though this approach has been in revealing the secret of organic change, it is bound to be, without the checks provided by paleontology, in some degree mythological and figurative.

Lacking adequate human materials, the attention of the Darwinists was necessarily confined to the depths of the Bornean and other tropical rain forests, partly because of the known great apes that haunted their depths, but also partly because of the feeling that other more human "anthropoids," true "missing links" might still be lurking thereabouts.

A typical expression of this view actually precedes the publication of the *Origin of Species.* It occurs in *Chambers' Journal* in 1856 and reveals once again the popular interest in the orang.[5] The anonymous author, after commenting that the animal may possibly be taught to speak, ventures a remark which reveals his concern with living missing links in the human phylogeny. "The Mias papan," he ventures, "may form only the external link of a chain, the other extremity of which lies hidden in the wild solitudes of Borneo. . . . We would therefore," he urges, "suggest to philosophers the desirableness of giving a new direction to their researches, and trying what may be done in the regions of the further east." Toward the close of the nineteenth century those regions would yield *Pithecanthropus,* but earlier it was still possible to seek for a living connection between man and the great anthropoids.

It may thus be seen that prior to the discovery of the Neanderthal cranium, and prior, also, to the publication of the *Origin of Species,* the rising interest in the geological history of the planet had done nothing but extend suspicions of man's antiquity and lead some individuals to anticipate a relationship between man and the great primates. If man was really an educated ape who had learned to speak and wear clothes, some degree of time must be assumed in the process, but this time element still did not seem to demand the assumption of a long chain of extinct ancestral hominids.

As evolutionary and geological speculation intensified past mid-century, Dr. Fühlrott and Dr. Schaaffhausen announced the discovery of the first human cranium recognizable as lying outside the known limits of human variation. Schaaffhausen must be accredited with a compete recognition of the skull as "due to a natural conformation, hitherto not known to exist even in the most barbarous races."[6] By a fortunate chance the discovery only slightly anticipated the publication of Darwin's and Wallace's evolutionary theories, and thus it drew more attention than would otherwise have been the case. Before long, argument would rage as to whether it was a genuine early link in the human phylogeny, yet at the time of Schaaffhausen's first report it is interesting to note that he merely expressed a suspicion that the skull might represent an individual member of one of those wild northern races spoken of by the Latin writers.

So far I have made mention of the hints and intimations of human primitiveness derived from the examination of living apes and their arrangement in a sequence of stages with existing man. It is not, therefore, without interest that when a cast of the Neanderthal calvarium reached London it was exhibited between the skull of a gorilla and two Negro crania. "Should this Neanderthal man prove to be an intermediate species between the Papuan and the gorilla, a great point of controversy would be gained by the transmutationists . . .", commented one observer.[7] Another anthropologist, Pruner-Bey, did not weigh the Negro resemblance so highly, but was fascinated by the close resemblance of the skull to that of existing Irishmen.[8]

"Part at least of the ancient memorials of Nature were written in a living language," Sir Charles Lyell once wisely observed. That key, that Rosetta stone to which Lyell refers is, of course, comparative anatomy. Without it we would be at a loss for a clue by which to understand the life of the past, or its relationship to ourselves. It is for this reason that Neanderthal man and his interpretation bulk so large in the mid-nineteenth century. He was, to pursue Sir Charles' simile, the first archaic human syllable caught echoing from the remoter past of man, yet still recognizably human. It was inevitable that his discovery should raise a storm of contention. Crania of similar rugosity were thought to be observed among the living peasantry of Europe, preferably among nationalities other than one's own.

There was ample justification for Rudolph Virchow's remark that after the Neanderthal discovery, skulls from peat bogs, from time levels known today to be Neolithic, began to be regarded as primitive. "They smelt out," he wryly remarks, "the scent of the ape." And, further, if it was pointed out that these crania from the old pile villages were capacious and modern, some of the more rabid evolutionists contended that these people "had more interstitial tissue than is now usual, and that in spite of the size of the brain, their nerve substance may have remained at a lower stage of development."[9] It was, to put the matter bluntly, a way of arguing for evolution by saying that even when ancient skulls had the appearance of *Homo sapiens* they had, to all intents and purposes, been stuffed with connective tissue, a polite euphemism for sawdust. Those who could castigate Virchow today for his opposition to the acceptance of the Neanderthal skull as a valid human fossil must remember that such excesses as represented by the above statements did nothing to increase the tolerance of the old master scientist.

Turning from the Continent to the British Isles, it may be asked what views were entertained by Darwin and Huxley. Darwin, always cautious, did not express himself early. Huxley, writing to Lyell in 1862, commented that "the Neanderthal skull may be described as a slightly exaggerated modification of one of the two types (and the lower) of Australian skulls."[10] Darwin, laboring upon *The Descent of Man* (1871), and drawing from Carter Blake before him, spoke of the enormous projecting canines of the La Naulette specimen. Nevertheless, he is on the whole wary, particularly because he was impressed with the capacity of the Neanderthal cranium. The Darwinians who had eyed microcephalic idiots as possible reversions to primitive man were actually not very happy or enthusiastic about the first Neanderthal discovery.

The massive supraorbital development attracted them, but the face was missing, a factor which made it impossible to be quite certain that the Neander skull was not within the variant range of the modern races. Some, like Carl Vogt, labeled the Neanderthal cranium as that of an idiot, but one has to realize that in the eyes of certain of these workers, and particularly Vogt, this does not destroy the value of the specimen as representing an early human stage of development.[11] Instead, this theory rather cleverly evades the issue of time. It is equivalent to saying: if the specimen is old it is "normal" for its period. If it should prove to be a comparatively recent burial it is "abnormal," but a remarkable case of reversion, nevertheless, to a primitive ancestor.[12]

The Western world throughout the last few centuries has tended to see itself in terms either of steps going up, or of steps coming down. Progress or degeneration has been a debate often influenced by the individual temperament of philosophers and perennially renewed under the impact of new systems of thought. The role played by the discovery of Neanderthal man cannot be entirely grasped without some reference to this aspect of the Western mind. Under the influence of Archbishop Whately, a strong belief in the conception that modern primitives and even archeological remains of past primitives represented a state of degradation from more civilized conditions was variously expressed by numerous writers at the mid-century and beyond.

"We have found nothing yet," comments H. B. Tristram, "to prove that the barbarous dwellers on the kitchen middens were not the wandering outcasts from the pre-existent civilization of the valleys of the Euphrates or the Nile, nor is there any chronological argument against it. Nor have we yet seen the traces of the barbaric epoch underlying the vestiges of the earliest civilization in its sites. Nor in the face of the relics of the Mississippi valley, of Central America, or of Mesopotamia, can we admit that there is no evidence before us of man relapsing from civilization."[13]

It would be impossible in a short space of time to explore fully the degeneration theory which, if pursued, could be shown to be related to the "Decay of Nature" concept of the Elizabethan era. As an argument to explain certain archeological facts, however, it re-emerges a little prior to the publication of the *Origin of Species* and at a time when the bone caves of Europe are just beginning to be probed. The revelations of the caverns were beginning to trouble religious orthodoxy. There can be no doubt that the emergence of the "degeneration" idea was the direct attempt of the spokesman for the traditional viewpoint to explain away the crude artifacts which, when found in Europe, hinted that remote and bestial beginnings might lie at the root of western culture.

Though the argument is no longer attractive, it was a persuasive alternative to evolutionism in the mid-century decades. There were serious debates, and some of the leading scholars of the period—men like the Duke of Argyll and Sir John Lubbock —arrayed themselves on opposite sides of the question. Archeology could supply the material for debate, but the truth was that its materials lent themselves readily to either the degeneration or the development hypothesis. Rudolf Schmid, a contemporary student of the period, confessed frankly: "Archeology, as a whole, seems to do no more than admit that its results can be incorporated into the theory of an origin of the human race through gradual development, *if this theory can be shown to be correct in some other way.*"[14] Schmid, in other words, recognized that something other than the approach through artifacts was needed to establish the reality of human evolution. That something was provided by the discovery of the Neanderthal skull. The skull and the La Naulette mandible of 1866 focused attention on the fact that the bone caves might contain more than simple artifacts and evidence of extended human antiquity;

they might provide conclusive proof as well that the human body itself had undergone the transfiguring touch of time.

It is true that the Darwinian passion for present-day atavisms obscured, for a time, this revelation, and that the skulls of idiots occupied on the demonstrator's platform a place between the skull of the orang and the Hottentot; but as the evidence from the caves slowly accumulated, this fashion passed. The Neanderthal discovery, coming, as it did, early enough so that translations and accounts appeared in England almost simultaneously with the opening of the evolutionary debate, insured that the skull would not be ignored or forgotten.

For a striking new fact to receive attention, it must fit into a theory, and that theory was immediately at hand. The earlier Neanderthal discovery at Gibraltar had had no such good fortune. Schaaffhausen's remarks about the wild tribes mentioned by the Romans would be quickly forgotten by their author, nor would it matter that the Neanderthal specimen would be challenged from every conceivable point of view. The real point lay in the world-wide, if hesitant reception of the fossil, in the interest it had stirred. Out of the bones would grow a new and specialized science, and as a consequence there would occur a slow fading of the degeneration hypothesis. The discovery would lead on, also, to the truly historical morphology which Darwin and his colleagues had not yet succeeded in extricating from the comparative morphology of the living which continued in some degree to dominate their epoch and their thinking. Nor did it dominate biologists alone. As late as 1905, a prominent American anthropologist, W. J. McGee, who had charge of the anthropological exhibits at the Louisiana Purchase Exposition in St. Louis, spoke as follows: "The next physical type chosen was the Ainu . . . their small stature . . . their use of the feet as manual adjuncts, their elongated arms and incurved hands, and their facility in climbing, approximate them to the quadrumanes and betoken a tree-climbing ancestry."[15]

Alfred Russel Wallace would, in his emphasis upon the shift of evolution to the mind, with its enormous latent possibilities, achieve, in the end, a partially effective compromise between the degenerationists and the evolutionists, for Wallace had a foot in both camps. He would explain the apparent slowness of human evolution as then understood, and, at the same time, in his meditations over the mysterious powers of the human mind, he would go far to abandon the linear biology which placed living races in a series of subordinate "fossil" levels, and which revolted many Christian thinkers.

Wallace, who unfortunately never wrote a book on the subject, probed deeper into the nature of man than any of the circle immediately around Darwin. Because in the end science has so thoroughly accepted them, we have not only forgotten their source but also forgotten how heretical some of his views were at the time they were uttered. First Wallace postulated an erect, small-brained bipedal stage of human development, followed by a second phase in which the human brain and cranium assumed their present size and form. Only with the present-day discovery of the Australopithecine man-apes is the early stage beginning to be documented. Second, he quickly saw that the complete fossil history of man might well be prolonged far beyond Pleistocene times, and that the big-brained men of the upper Pleistocene, who were at that time troubling the evolutionists, need not be regarded as an effective argument against the reality of the human transformation. Rather, the scientists must cease confusing living races with grades or levels on the evolutionary scale of the past—something which was at that time exceedingly common. Natives were often described as apelike in appearance, and there was a strong unconscious tendency to see the whole story of human evolution revealed in a sequence from such existing apes as the gorilla through the "lower" living races to Caucasian man.

It is just here that Wallace's deep humanity and long experience with primitive peoples reveal with particular clarity his more sophisticated approach to the subject. At a time when many naive observers were comparing the languages of primitives to the chattering of apes or monkeys, Wallace stated unequivocally: "Among the lowest savages with the least copious vocabularies the capacity of uttering a variety of distinct articulate sounds, and of applying to them an almost infinite amount of modulation and inflection, is not in any way inferior to that of the higher races."

Wallace believed that, however the fact had come about, all the existing races were mentally pretty much equal, having attained their *Homo sapiens* status a long time ago. Why human beings scattered over such remote distances should be so similar in intellectual capacity, while at the same time varying so vastly in their technological achievements, he did not profess to know. Of one thing, however, Wallace felt certain. Man, even savage man, possessed latent mental powers—ability to understand and produce music, mathematics, art—which were not accountable for, Wallace felt, in terms of the simple utilitarian struggle for existence as portrayed by the Darwinists.

If an attempt is made to summarize briefly the separate elements of human thinking which were in some degree altered by the discovery at the Neanderthal cave, they may be listed as follows:

First, the discovery aroused and maintained interest for a number of reasons. Anatomically, the massive character of the orbital ridges and the unusual conformation of the vault led to suspicions that its type was extinct in Europe. Also, and most importantly from the speculative standpoint, the skull came to serious attention almost simultaneously with the developing Darwinian controversy. Seized upon by some of the more avid evolutionists as a proof of man's descent from the great anthropoids, it was interpreted in terms of the taxonomical ladder which had been utilized by the transformationists of the first part of the century. Its capacious brain-case did not prevent it from being labeled as a brute, and its characters were so transformed that, without the slightest basis in fact, it was described as possessing huge projecting canines and "an appearance in the highest degree hideous and ferocious." This ferocity, of course, represents a distorted Victorian emphasis upon the struggle for existence in its more lurid forms.

Second, as the number of big-brained primitives in the upper Pleistocene became slowly established, the Darwinists were forced to reassess the time involved in the human transformation, and to extend it.

Third, the most astute researchers were led to consider the possibility that the earth, rather than the jungle or the insane hospital, might contain the clue to man's evolutionary past.

Fourth, the degeneration theory, successfully advocated by Archbishop Whately and his followers as long as they had only the archeologist to face, began to lose its efficacy once it could be shown that the caves of Europe contained actual organic traces of men more primitive in physical type than men of the present day.

Finally, what lesson has this discovery to teach us today —over one hundred years after the quarrymen sank their shovels in that little enclosed grotto in the gorge of the Neander? Man, irrespective of whether he is a theologian or a scientist, has a strong tendency to see what he hopes to see. The nineteenth-century evolutionists—those who accepted the Neanderthal specimen—saw a savage animal whose skull contours reminded them, paradoxically, of the Negro, the Mongol, the Hottentot or the Irishman—all peoples at that time economically depressed. The scholars tended to cling to a linear phylogenetic line in which each living race represented a frozen step on the way to the emergence of the civilized Caucasian. We smile at this today, yet it may be wondered whether our present frequent tendency to arrange our varied assemblage of human fossils in a single unilinear phylogeny may not threaten to lead to a historical simplification as rigid and potentially wrong as we now know the unilinear taxonomical approach through the living races to have been. After all, it is known paleontologically that families have proliferated and later contracted as a single type has achieved ecological dominance.

When we consider this creature of "brute benightedness" and "gorilloid ferocity," as most of those who peered into that dark skull vault chose to interpret what they saw there, it should be remembered what was finally revealed at the little French cave near La Chapelle aux Saints in 1908. Here, across millennia of time, a very moving spectacle can be observed. For these men whose brains were locked in a skull reminiscent of the ape, these men whom serious scientists had contended to possess no thoughts beyond those of the brute had laid down their dead in grief.

Massive flint-hardened hands had shaped a sepulcher and placed flat stones to guard the dead man's head. A haunch of meat had been left to aid the dead man's journey. Worked flints, a little treasure of the human dawn, had been poured lovingly into the grave. And down the untold centuries the message has come without words: "We too were human, we too suffered, we too believed that the grave is not the end. We too, whose faces affright you now, knew human agony and human love."

It is important to consider that across fifty thousand years nothing has changed or altered in that act. It is the human gesture by which we know a man though he looks out upon us under a brow suggestive of the ape. If, in another fifty thousand years, man can still weep, we will know humanity is safe. This is all we need to ask about the onrush of the scientific age.

The Intellectual Antecedents of *The Descent of Man*

I

DARWIN HAD cautiously avoided direct references to man in the *Origin of Species.* But twelve years later, after its triumph was assured, he published a study of human evolution entitled *The Descent of Man.* Meanwhile, Lyell had written a cautious volume (1863) dealing with human antiquity, which disappointed Darwin; Huxley had treated the anatomical aspects of man's relationship to the primates in *Zoological Evidences as to Man's Place in Nature* (1863); and Carl Vogt had contributed his *Lectures on Man* (1864). By 1871, then, the subject was scarcely new; in fact, it could not have been new since the time of the *Origin,* the implications of which were perfectly plain. Nevertheless, the world wanted to hear what the author of the *Origin* had to say on the evolution of man. St. George Mivart observed in *Nature* the year the *Descent* was published that "already there is a perceptible increase in the visitors to the monkey house."[1]

Huxley's brief work was written with wonderful clarity and directness. By contrast, *The Descent of Man* has some of the labored and inchoate quality of Darwin's overfull folios of data. It is contradictory in spots, as though the author had simply poured his notes together and never fully read the completed manuscript to make sure it was an organic whole. Some critics, including Wallace, have remarked that the *Descent* is two books, two subjects, mixed together. Some two-thirds of the work deals with sexual selection, and only a small proportion of the latter material has a bearing on man.

Another of its defects is Darwin's failure to distinguish consistently between biological inheritance and cultural influences upon the behavior and evolution of human beings. In this, of course, Darwin was making a mistake common to biologists of the time. The worlds of the human paleontologist and the cultural anthropologist were scarcely cleared of morning mist. No one had yet unearthed any clear fossils of early man. A student of evolution had to content himself largely with tracing morphological similarities between living man and the great apes. This left considerable room for speculation as to the precise nature of the human ancestors. It is not surprising that they were occasionally visualized as gorilloid beasts with huge canine teeth, nor that Darwin wavered between this and gentler interpretations.

An honest biographer must record the fact that man was not Darwin's best subject. In the words of a nineteenth-century critic, his "was a world of insects and pigeons, apes and curious plants, but man as he exists, had no place in it." Allowing for the hyperbole of this religious opponent, it is nonetheless probable that Darwin did derive more sheer delight from writing his book on earthworms than from any amount of contemplation of a creature who could talk back and who was apt stubbornly to hold ill-founded opinions. In any case, no man afflicted with a weak stomach and insomnia has any business investigating his own kind. At least it is best to wait until they have undergone the petrification incident to becoming part of a geological stratum.

Darwin knew this. He had fled London to work in peace. When he dealt with the timid gropings of climbing plants, the intricacies of orchids or of the carnivorous sundew, he was not bedeviled by metaphysicians, by talk of ethics, morals or the nature of religion. Darwin did not wish to leave man an exception to his system, but he was content to consider man simply as a part of that vast, sprawling, endlessly ramifying ferment called "life." The rest of him could be left to the philosophers. "I have often," he once complained to a friend, "been made wroth (even by Lyell) at the confidence with which people speak of the introduction of man, as if they had seen him walk on the stage and as if in a geological sense it was more important than the entry of any other mammifer."

Man is a unique and troubling expression of the evolutionary forces. In nineteenth-century terms it was easier to explain his rise anatomically than to account for some of his distinctly human traits such as language and culture. The tools for the precise examination of the human story were yet to be forged. Some remain elusive even to the minds of this century, some are just appearing in the new primatology and ethology. But the world awaited Darwin's word on man; the author of the *Origin* could scarcely be expected to escape what, after all, was the heart of the controversy he had precipitated. Tired and shaky, he dipped his pen and wrote as though the last sentence of the *Origin* still lingered in his mind: "We are not here concerned with hopes and fears, only with the truth as far as our reason allows us to discover it." It was well that he added the caveat. He had never been a truculent writer like some of his younger followers. Unlike them he had come too far alone.

There is the story of a visit, a last visit, that he paid to his boyhood home in Shrewsbury while he was engaged upon *The Descent of Man.* The later owner was gracious but untactful. He paced, talking expansively, from room to room with his weary visitor who wanted only silence and dreams. As Darwin descended the steep slope towards the town he remarked uncomfortably to his sister Henrietta: "If I had been left alone in that greenhouse for five minutes, I know I should have been able to see my father in his wheel chair. . . ."

In a way the little episode epitomizes Darwin's role in the intellectual life of the century. He was the dreamer destined to destroy simple belief, yet looking nostalgically backward into empty rooms and hearing silenced voices. Perhaps no man in science heard more such voices or listened longer at the doorways of the past. "What an enormous sum Darwin availed himself of and reinvested," once exclaimed John Burroughs.[2] The literary naturalist had been greatly perceptive, but few of Darwin's contemporaries had observed the reluctance with which the fate-scarred man pursued his destiny. He appears a figure of Delphic ambiguity, an ambiguity that occasionally erupts in his own phrases. In short, Charles Darwin was haunted, and one of the subjects by which he was unconsciously haunted was something that the nineteenth century called "natural law."

II

The physicist Max Born once remarked that science progresses through a series of intersecting and confusing alleys rather than down the broad and ever-widening highway that is the illusion fostered by our technological successes. Through these obscure alleys slip elusive shadows of thought that sometimes vanish only to reappear in a succeeding century. Other similar shadows grow large, as though incorporating the thought substance of an entire generation, but then undergo transmutations never envisaged when they first appeared. Darwin's world, the world of the Victorian era, like all centuries, had its roots in the metaphysics of the past. It was the inheritor of Galileo's thought and that of the great intervening mathematicians and astronomers, culminating in Sir Isaac Newton in the seventeenth century. The thought of these discoverers, devout though they were, was to give rise to a conception of natural law that would, by degrees, dismiss the Deity to whom they avowed allegiance and fix upon Western man a determinism so severe that it gave rise to a literature of enormous melancholy, of which Hardy's epic poem *The Dynasts* is perhaps the last grand example:

> *Things mechanized*
> *By coils and pivots set to foreframed codes . . .*

Natural law has a history extending into antiquity but, like the word "nature" itself, it has assumed more than one meaning in the course of time. Only in the sixteenth and seventeenth centuries, and paralleling the rise of science, did the term come to assume supreme importance in Western thinking. Prior to that time God's reign over nature did not hamper Him from making miraculous exceptions, staying the sun, sending warning comets to attest great events, or otherwise participating actively in His world. Whether, as some have contended, the rise of immutable natural law has some correlation with the growth of centralized authority in the capitalist state or whether its rise merely represents a growing scientific sophistication, is difficult to analyze.[3] As in all complex historical situations, precisely identifiable causes may be difficult to distinguish from accidental but correlated events. To attribute the recognition of unvarying rules of nature purely to the emergence of state authority seems a somewhat simplistic Marxist invention.

The deterministic machine did not arise immediately among the natural philosophers; instead it evolved. Newton, whose mathematical triumphs clinched the direction of science for the next two centuries, was himself a religious man. He drew his conception of God as a nonmaterial being extended throughout space and time largely from Robert Boyle, who in turn had been influenced by the Cambridge Platonist Henry More. Newton, indeed, remarked that "a god without dominion, providence, and final causes is nothing else but Fate and Nature." To that position certain of the eighteenth-century materialists would paradoxically proceed, dropping along the way the three Newtonian attributes of divinity. In the meantime, it may be well to inquire just how the unquestionably devout Newton had succeeded in creating a world machine that, in other hands, would negate the necessity of Providential control.

First of all, Newton possessed a strong empirical bent linked with mathematical genius of a very high order. He moved in a world of primary qualities from which the organic world had been largely abstracted. Moreover, the Newtonian system of nature treats of the existent world, not the emergent world of novelty then unglimpsed. It is essentially a world of masses of varying distance and size affecting other masses in a calculable, if invisible, way.

Galileo distinguished between primary qualities such as the laws governing the movement of bodies and those secondary qualities such as light, color and sound received by the human senses. This honest attempt to refine our conceptions produced by degrees a scientific distrust in anything at all subjective or unmeasurable by mathematical formulae. Galileo had termed the subjective, secondary qualities "mere words." Philosophers enraptured by the night sky no longer were stirred by the light and color of the living world. They had come to mistrust every sense but the mathematical. To obtain an ordered universe subject to one unifying system of law, they were willing to enter the cold, dark, soundless world of the void and remain there mentally. Not for nothing has Basil Willey commented that when the Romantic movement began a hundred years after Newton, "several of the leading poets attacked science for having killed the universe and turned man into a reasoning machine."[4]

The word "machine," with God cast in the role of the Great Mechanic, epitomized seventeenth-century science, until Newton's feats of pure reasoning subjected the entire universe to one unifying system of law. There remained no longer a divine Empyrean, a celestial sphere beyond human ability to penetrate. Instead, earth and sky had been drawn into the same sustaining web of mathematics. Like Darwin, however, Newton was more cautious than some of his associates. In his *Universal Arithmetic* he expressed doubts that some problems could be handled mathematically at all—a heresy, says E. A. Burtt, to men like Galileo or Descartes.[5] Nevertheless, it is the world of the great clock that took precedence in seventeenth-century thinking: a world of absolute law, of silently spinning wheels set into operation by a remote cosmic intelligence no longer concerned to intervene directly in the workings of the world. Providence loses its original interpretation; at best it is confined, one might say, to swinging the pendulum that started the clock. Or again, it may occasionally be needed to make slight corrections in the cosmic machinery. Man exists, in essence, merely to worship the machine—a gigantic version of Boyle's Strasbourg clock, which, once started, requires no artificer. The world of the secondary qualities, the world of color, of organic activity, no longer mattered. At best the latter came to be treated as a world of smaller machines marching to the laws of the greater.

The providence so long believed in by the Christian world,

the providence of personal divine intervention on behalf of individuals, the concern for the falling sparrow, was slowly being abandoned for a realm of unchanging, inflexible law laid down by an awe-inspiring but inconceivably distant deity. Providence had become merely a concealed name for the existent universe, its preservation and suspension in time. Newton and most of his colleagues were religious men imbued with wonder at the regularities of the cosmos. Where the worship of this regularity was to lead would become manifest only in another century.

III

Observing the course of that great instrument, the providential clock, elaborated in the mind of Newton and more particularly in that of his less restrained eighteenth-century followers, one is struck by the rapidity with which over two centuries the role of deity recedes and is replaced by causal law, a law so strict that a mind capable of positioning all the particles in the universe could foretell their movements and predict the entire future.

The bright world of secondary qualities was fading like an illusion. In the cold and darkness an irrelevant spectator, man, wandered among the flickering fires of the new industrialism. He had entered that "City of Dreadful Night" which James Thomson was to describe so graphically:

I find no hint throughout the Universe
Of good or ill, of blessing or of curse,
I find alone Necessity Supreme;
With infinite Mystery, abysmal, dark
Unlighted ever by the faintest spark
For us the flittering shadows of a dream.

As early as 1833 Carlyle wrote of the cosmos as a "steam engine rolling on in its dead indifference, to grind me limb from limb." Haeckel spoke of the "absolute dominion of the eternal iron laws of the universe." W. K. Clifford mourned with simple sadness that "the Great Companion is dead." Finally, as an epitome of this thinking Kingsley had protested "tell us not that the world is governed by universal Law; the news is not comfortable, but simply horrible."[6] Science, in breaking out of the closed world of the past, had made an orphan of humanity, an orphan in whom free will was well-nigh extinguished in favor of total mechanism. What might be called the providential machine of Newton, the machine governed and sustained by an increasingly remote divinity, had been insensibly altered by the nineteenth century into a world of inflexible law, of Calvinistic gloom and necessity. It constituted the intellectual world into which Charles Darwin was born.

Yet in observing major trends of thought one must take care to allow for the loose ends and diverse tendencies that constitute the thinking of any century. It is all too easy in retrospect to schematize beyond reality. No satisfactory study of what natural law meant to the nineteenth century exists in the literature. The era, however, contains harsh expressions of its significance, expressions which, in some instances, conceal a lurking religiosity. This trend is illustrated by attacks on the early evolutionists. The astronomer John Herschel, for example, employed some rather dubious reasoning with which to castigate Robert Chambers.[7] Here the reign of natural law is obviously used as a bulwark against any acceptance of emergent novelty.

"Law" and "Baconian induction" are the conservative catchwords of the early part of the nineteenth century. The one rigidly determines the possible, as if man had received a full revelation on law; the second, induction, is frequently called upon to combat uncomfortable speculation. The offender was accused of failing to accumulate sufficient facts to document his hypothesis. The conservative critic was thus placed in a strong position behind a great name. It is he who constitutes the sole judge of "fact." It is obvious that Darwin felt the weight of this academic subterfuge when he protested ingenuously, "I worked by true Baconian induction. . . ." It was dangerous to suggest one might have entertained an insight or a supposition before one examined "facts." Robert Chambers had been so heavily bombarded by critics of this persuasion that Darwin's wariness was greatly heightened. He had studied both Chambers and his critics. Thus Darwin had the advantage of observing what was likely to be a preview of his own later experience. It led him, no doubt, into some further procrastination, but it doubtless increased his determination to avoid hasty generalization, fond though he might privately be of speculation.

The full scope of the history of natural selection affords an entertaining example of how, in biology, the balanced, Newtonian world machine was later to give rise to a new conception of the struggle for existence. Historians of science have largely neglected the fact that, like the opposed faces on a coin, natural selection is actually made up of two superficially contradictory theorems, one of which was adopted before the other. The earlier theorem bore appellations other than natural selection. When the latter phrase was introduced by Darwin, the earlier usages faded away almost entirely without recognition that they bore an historical connection to Darwin's newly introduced terminology. The whole episode is revelatory of the importance of semantics in the shifts and permutations of scientific thought.

By the late eighteenth century the struggle for existence had been recognized by a number of writers, the great majority of whom were not evolutionists. They observed in "struggle" a providentially imposed system, a "machine," as it were, for controlling animal numbers. Terms such as "pruning," "policing," and "natural government" indicate a recognition of population control and stabilization within particular environments. The cutting edge of selection is fully delineated in expressions such as "keeping the species up to par." "Natural government" is seen by most of these writers as a completely conservative mechanism balancing one species off against another in a system of checks and balances whose aberrations are, in the long run, as correctable as the Newtonian scheme of the heavens. Nature, in short, is still *existent* nature. It contains no room for emergent novelty; it does not promote change.

Or does it? In the late eighteenth and early nineteenth centuries, a period marked by the rise of the new industrial cities, and of expanding urban populations that had to be fed and clothed from the countryside, a class of animal inventors or innovators arose who introduced new and highly successful breeds of sheep and cattle. One of the more successful of these rule-of-thumb breeders was Robert Bakewell, whose career attracted the later attention of Darwin.

The significance of these events that took place originally outside the domain of theoretical science lay in the fact that the farm breeder, practicing artificial selection, was drawing out of his living plants or animals changes in form or texture that had existed as mere potential within the species. This novelty was all directed to practical ends by practical men. That nature might have provided such hidden variability for other than human aims went unremarked. The standard interpretation of the Christian time-scale was too short to allow for such massive transformations of life as those that occupy the modern evolutionist. Nevertheless, from the time of eighteenth-century Newtonian natural government to Darwin's recognition of natural selection as an effective explanation of organic diversity and change, a slow philosophical shift had been in progress. A purely conservative organic paradigm was in the process of giving way to one that could embrace, in the term "natural selection," both the conservative and the novel aspects of the struggle for existence.

It is even more remarkable that in the multitudinous reviews of the *Origin of Species* following its appearance in 1859, only one solitary commentator, H. B. Tristram, to my knowledge, ever glimpsed the relation of the earlier paradigm to the later. He wrote, piercing the semantic barrier that by 1866 enclosed the new biology: "As to the struggle for existence, the facts adduced by Mr. Darwin would equally harmonize with the theory that the struggle tends to *conserve* the type, as with that which maintains that it tends to *change* the type."[8]

In these words Tristram had correctly indicated the seeming paradox of the two major facets of natural selection. Both are true. As to which prevails on any given occasion, only the contingencies locked in the genes and the opportunities presented by the outside environment may determine whether a given phyletic response will partake of static or novel elements. The emergent world of change that had arisen like a phantom between the fingers of the gardener and stock breeder, the world glimpsed by Darwin in the Galápagos and in equally long meditations over the pages of agricultural books, was about to overthrow the conception of eternal balance.

Immediate nature was about to be replaced with something more intangible, a nature having to do with the latent and the possible. Such a nature of pure lurking potential was beginning to hover like a mist around what had previously been understood to be a single creative act of divine will. Statistical probability, later to be called Darwin's "empire of accident," threatened to overthrow the iron causality embedded in nineteenth-century physics. It is perhaps little cause for surprise that, on the whole, physicists like Lord Kelvin and astronomers like Herschel viewed the new movement with distaste. Ironically, natural selection, by displacing man from the center of cosmic attention, would readily enhance the gloom that already lingered over a world from which personal providence was departing. William Blake's "Satanic mill" of the industrial city was beginning to merge with the Satanic machine that had become the universe.

IV

Darwin was born in a century which regarded the "laws" of nature as imbued with a kind of structural finality, an integral determinism, which it was the scientists' duty to describe. Biology and geology were still occupied largely in the classification and description of facts. Fossils, mountain-building and erosion, animal and plant distributions, were beginning to tell man something of the earth's antiquity but only in the face of much emotional bias. It was not easy to perceive, behind the well-ordered world of the present, an extended and coalescing chain of phylogenies all pointing downward into unspeakable gulfs of time.

More germane to the issue, perhaps, was the strain that this whole achievement placed upon existing regularities. To dig into the past and discover immanence, to examine the present and find it laden with a novel organic future, meant that if one was to avoid charges of violating the persisting uniformity of the world one had better be prepared to elucidate the laws governing these strange phenomena. It is to Darwin's credit that he made the attempt, but it is also evident from his language and his hesitations that he was sore beset as to how to describe the world of contingency into which he had wandered.

Variation was a fact of nature; that he knew, as had the plant and animal experimenters before him. He lacked all knowledge of its cause but clearly recognized that it supplied the raw materials upon which selection operated. If we go back to the earliest years of Darwin's efforts we find that in a letter addressed to the geologist Lyell in 1837 when, as we know, he was first finding his way into the subject of natural selection, he speaks as follows: "Notebook after notebook has been filled with facts which begin to group themselves *clearly* under sub-laws."[9]

"Sub-laws" is the key word. It reveals Darwin dealing with an unruly and emergent novelty, seeking, after the manner of the period, to fit it into the inflexible natural law of his contemporary world.

Again, writing privately to Hooker in 1849, he reiterated his concern with the unpredictable variability which in some slight degree is everywhere to be found in the organic world.[10] "What the devil," he puzzles as late as 1859, "determines each particular variation? What makes a tuft of feathers come on a cock's head, or moss on a moss rose?" Darwin is here struggling with two separate but related problems, the first of which (the frequency of variability) he chose to underplay in the first edition of the *Origin of Species.* It was difficult to see how one captured under "natural law" what seemed to leap into existence with a fine disregard of anything but pure chance. In the same year he confesses, baffled, to Joseph Hooker, "I believe . . . perhaps a dozen distinct laws are all struggling against each other in every variation which ever arises."[11] Laws Darwin seemed determined to pursue, but they continued to elude his best efforts. To explain coordinated evolutionary structures he was forced to fall back upon "mysterious laws" of correlation. Again, he maintained to Lyell that species formation "has hitherto been viewed as beyond law . . .," the implication being that for the first time he is bringing scientific law into a subject previously confined to a "theological phase of development."[12]

Robert Chambers in the 1840s had been taken severely to

task for proposing in *Vestiges of the Natural History of Creation* a "law of development," which, as noted, had received the scornful attention of John Herschel. Darwin had, very obviously, taken note of this assault, and was determined, both logically and diplomatically, not to fall into the astronomical trap of equating his discovery, mathematically so indefinable, with a law like that which Newton had applied to gravity.

It is apparent from Darwin's correspondence that he continued to regard the analogy with gravity in a slightly wistful manner; but, in the end, he was content to term natural selection as, at various times, a hypothesis, a theory, or a broad metaphor.[13] Writing to Asa Gray in 1859 he ventured to explain it "as a geologist does the word denudation—for an agent, expressing the result of several combined actions."[14] Similarly, Darwin elsewhere indicated that the fall of a handful of feathers to the ground under the pull of natural law, that is, gravity, was easier to chart than the actions and reactions of the innumerable creatures and plants determining the mode of change among themselves. Darwin's caution, while it does credit to his philosophic sense, was to prove hopeless in terms of evading the sharp-tongued Herschel, who promptly accused him of promulgating "the law of higgledy-piggledy."

With that phrase the Empire of Accident, in spite of Darwin's wish to abide by the nineteenth century's conception of order, had indeed arrived. In the first edition of the *Origin*, Darwin observed diffidently that "variations useful in some way to each being in the great and complex battle of life should sometimes occur in the course of *thousands of generations.*" (L. E.). His colleague Wallace was quick to observe that in making this remark, Darwin was playing into the hands of his opponents. Variation, in Wallace's view, was omnipresent, and favorable mutations by no means confined to distantly isolated and rare episodes.[15]

In later editions of the *Origin* Darwin quietly accepted Wallace's criticism and incorporated it into his work. The world of universal law, whether that ordered by a remote impersonal providence, or maintained by the "dreaming dark dumb thing" of Hardy's despair, was slowly giving way to something else.

The Satanic machine was about to be replaced by the novelties of the probability machine to which, later on, even nineteenth-century physics was to succumb. Somewhere a connection had to be made between mathematical chance as it was conceived by the great gamesmen and gamblers and *visible* biological novelty. Had the idea lurked hidden in the work of the eighteenth-century breeders? Perhaps it had.

Long ago when Darwin was still a youth aboard the *Beagle*, the naturalist Robert Mudie, faithful to his century, had written: "There is a law that maintains the species." Scarcely had he made this assertion before he was busy explaining that all cultivated plants or animals were more or less monsters and that of the appearance of their parentage we knew little or nothing. Even of wild forms he ends by hinting ambiguously of the emergence of species "altogether new." Finally he verges on complete heresy. "There is something," he almost whispers, "of the same kind in human beings."[16]

It may thus be seen that the wild, the fortuitous, the gambler's throw invisible in the germ cell, did not go entirely unnoted by perceptive naturalists prior to Darwin. In fact, it was upon their observations that he was to draw. Yearn as he might for law, for order, attribute as he would to unknown causes the tuft on a cock's head or the variant pattern on a turtle's shell, the secret escaped him. His contemporary Mendel, who discovered a vital part of the key to change, would die unrecognized. The laws that Darwin sought would never be laws in the sense he had dreamed. They would turn at best into statistical regularities, descriptions sometimes as elusive and difficult to master as the feathers he had described rocking downward under the unknown force of gravity. The prescriptive power of the Newtonian machine was imperceptibly fading.

"If science is an account of the lawfulness of nature," Spencer Brown, the mathematician, once observed, "then history is an account of its chaos."[17] In penetrating the past of the organic realm, Darwin and Wallace had been forced to sacrifice existent nature, nineteenth-century nature, and indeed man himself, for a wilderness of subjective possibilities which, in turn, led out of the present into a world of "indefinite departure." Nature, the twentieth century was to discover, would no longer be the old visible "lawful" nature of field and forest, the nature which many Victorians regarded as the direct expression of God's will.[18] Even the physicist, toward the end of the century, was to discover that his supposedly stable elements were capable of ghostly transmutation. The world beneath the atom did not follow the order of the macroscopic universe in which life moved and had its being. As C. J. Schneer puts it, "the natural law that prevails on the level of classical physics is simply an expression of the probabilities that prevail on this deeper, more fundamental level of experience."[19] In the words of J. Bronowski, chance is given a kind of order—the order with which probability theory operates.[20]

Darwin's critics, who frequently misunderstood him as believing that *any* living thing, moss or carrot or cabbage, could be transformed into anything else—the law of higgledy-piggledy—were quite wrong. Darwin, though he had no word for it, was dealing with what now could be called historical probability, a phylogenetic course which might wander within limits but which had a set of limiting parameters determined by past mutation and selection. Historical probability, in other words chance-accumulated and selected variation, will determine the viability of new mutations.

Nevertheless, if we allow for this stricture, we must still accept W. Heisenberg's observation that "these may be biological processes . . . in which large-scale events are set off by processes in individual atoms; this would appear to be the case particularly in the mutation of genes during hereditary processes."[21] Thus the prescriptive law as to what may be finally attenuates, thins and tends to vanish. Darwin, to a degree, yearned wistfully for the kind of law that would satisfy the men of his century. He did not find it. Perhaps in the end, however, we of this generation can look upon his efforts and admit as modern physics has had to admit, in the words of Chesterton, that "a thing cannot be completely wonderful so long as it remains sensible." Perhaps this was what kept Darwin at his interminable labors even when he strove wistfully to produce a law that would satisfy his antagonist, Lord Kelvin, whose truly Satanic mathematics would have allowed life and the sun but a fleeting moment of existence. For in nineteenth-century terms, he could prove, as Darwin could not, the mathematical reality of his conclusions.

V

I intimated in the beginning of this essay something of Darwin's original dilemma: namely, the problem of how organic evolution and, above all, human transmutation, could be interjected into a scheme of natural law which concerned only the existing Newtonian world. By force of logic and the geological time extensions afforded by the work of Hutton, Lyell and others, his views triumphed. It is equally clear, however, that the philosophical world which surrounded him and whose tenets he embraced haunted his studies to the end. "I am inclined," he wrote to Asa Gray in 1860, "to look at everything as resulting from designed laws, with the details, whether good or bad, left to the working out of chance. Not," he adds significantly, "that this notion *at all* satisfies me." Darwin's mind, troubled and insecure to the end, was wavering and rocking like autumn thistledown in its descent.

By 1865 Francis Galton, Darwin's cousin and an early worker in mathematical psychology, was intrigued with the concept that the moral and religious sentiments had been evolved through natural selection from the herd instincts of man's forerunners.[22] Huxley, in one of his lectures to working men in 1862, had perhaps been the first person to emphasize a relation between natural selection and morality.[23] It is very evident from K. F. Gantz's useful and too little known survey, as well as the equally valuable account of James Tufts, that once man had been stripped of his place in the Biblical conception of creation, the question of his moral nature and its origins was bound to become an object of debate. Some would point to man's powers of speech and his ethical nature as evidence that, whatever might be true of the rest of creation, man's role and origin upon the world stage must be different from that of his animal associates. Others accepted the concept of man's rise from the brute world with such enthusiasm that they rather forgot that man had risen at all.

The ancient quarrel that could be said to have arisen outside of Eden's gate, between Cain and his brother Abel, simply continued, as it continues today, in an evolutionary guise. As Tufts observed over fifty years ago:

> The striking thing . . . is the discredit which has now fallen upon the natural. One school of writers, indeed, maintains the rational and social nature of man, and the rational laws of cosmic nature, but the most striking evolutionary theories . . . conceive nature as the realm where force and the instinct for self-preservation hold sway. This was doubtless due largely to the theological dualism between the "natural" man born in sin, totally depraved with no good instincts, and the spiritual man who must needs be "born again," regenerated by special divine grace, before he would be just or good.[24]

This ancient quarrel still persists in the clash between Robert Ardrey and his opponents. It might be observed, in turn, that, although neither party realizes it, the controversy reflects the two opposing facets of natural selection as I have earlier described them. The conservative paradigm might be said to contain the party of "natural" man, while the liberal or emergent paradigm may well reflect the faith of others in man's possible perfectibility and potential mastery of his baser nature. The reductionist school here opposes the party of the future. The latter group, of course, have the history of man's rise to adduce as hopeful evidence, but the conservatives confront our optimism with the challenge that man clings to his midnight past. There the argument must rest in a time still so close to the forest and the ice. It might even be argued that the conservative school of thought has a lingering affection for the deterministic machine of nineteenth-century law and that its opponents place greater faith in the continued operations of the probability machine that called man into being.

Speculation about man's place in the natural world became a world-wide concern after 1859. Had Darwin never written again, books and periodicals would have continued to pursue the subject of man's moral nature as well as the evolutionary patchwork revealed in his body. It was regarded as right and appropriate that the first spokesman of evolution should speak about man. By 1871 ten years of accumulated data was available from a variety of sources, including Edward Tylor's *Researches into the Early History of Mankind.* Darwin drew heavily upon this literature as well as upon his own store of knowledge. In the sections of the book treating of man, Darwin compressed a sizeable amount of scattered speculation into a theory of naturalistic ethics.

One can state with surety about the body of Darwin's work that he thrice maintained continuities where the world had seen discontinuities; once in the origin of species, once in ethics, and once in language. The last two propositions he had defended in *The Descent of Man.* For these seeming great breaks he had argued the reality of multitudinous minute discontinuities, variations in the place of the great *saltus.* To a discriminating eye the little quantum jumps that accumulate by degrees into enormous novelties such as man in no way detract from the almost miraculous quality of the living world.

It would be foolish to expect, in a day when even Darwin made uses of anecdotal literature relating to these matters, that, in the anxiety to demonstrate continuities of development, parrots might not speak a trifle too intelligently, or use inheritance to promote speech or other cultural habits in a way no longer acceptable to the anthropologist. These are small matters which do not basically affect the greater achievement: the great dream and the realization of the human pathway through time.

Philosophically Darwin may have removed us from a privileged position in the universe but he brought us, at the same time, back into the web of life. He showed us by implication our relationship with and dependence on other organic beings and thus contributed indirectly to the rise of the modern conservation movement. Old maids and cats and bees and clover prepared us as children for the interlinked net we are now struggling to preserve.

Darwin did not realize as did Freud the vestigial remnants

still lurking below the conscious mind, nor, on the other hand, did he fully grasp that if ethics in some fashion arose from original primate sociality, it has been lifted beyond the purely utilitarian. In an excellent study of evolutionary moral theory, W. S. Quillian, Jr., observes that though one may start with herd instinct one ends with a value, a value that expresses a spirituality which is an end in itself.[25] The moral sentiments have therefore passed, or are passing, beyond the possibility of being totally formulated in organic terms. A dialogue with ourselves has begun. Man, as Darwin intimated, is a self-examining creature.

The final step was made by the master himself: man partook of the rest of creation. In A. E. Housman's words, he had blown hither "from the twelve-winded sky," tumbled about in organs and in mind by the working forces of elemental chaos, prodded onward by a "law" which, if it was a law, provided no sure direction, a "law" which to Darwin's mind brooded over the desolate landscape of accident. But the parameters were there, narrowing upon man for the next possible leap either to oblivion or glory. Man had become self-conscious, self-measuring, perhaps the only creature in the cosmos to be so. In some strange sense the creature is already transcendent. Perhaps some few of his kind have already surpassed him. Did not Einstein, with his twentieth-century mind that ranged as far into the universe as Darwin's, once remark perceptively, "It is far from right to say that nature always plays the same game"?

The Time of Man

I

IT IS A CURIOUS THOUGHT that as I sit down to write this essay on the history of our species, I do so in the heightened consciousness that it may never be published: a holocaust may overtake it. Tomorrow I may lie under tons of rubble, precipitated into the street along with the paper on which this history is scribbled. Over the whole earth—this infinitely small globe that possesses all we know of sunshine and bird song—an unfamiliar blight is creeping: man—man, who has become at last a planetary disease and who would, if his technology yet permitted, pass this infection to another star. If I write this history in brief compass it is because, on the scale of the universe, it is but an instant, shot with individual glory and unimaginable shame. Man is the only infinitely corruptible as well as infinitely perfectible animal.

The story I here record contains many gaps and few names. Most of what has gone into the making of man is as nameless as the nothing from which he sprang and into which, by his own hand, he threatens to subside. He has wandered unclothed through earth's long interglacial summers; he has huddled before fires in equally millennial winters. He has mated and fought for a bare existence like earth's other creatures. Unlike these others, however, he has clothed his dreams in magic that slowly became science—the science that was to bring him all things. And so, because of the dark twist in his mind, it has; it has brought him even unto death.

This story, then, is written in a kind of fatal haste—to be read by whom, I wonder? Or does it matter? Man is the talking animal. I shall talk to myself if necessary. I feel my own face before the mirror, touch the risen brain vault over the gray layer of cells that has brought us to this destiny. I lift up my hands under the light. There is no fur, black upon them, any longer.

For a moment I wish there were—for a moment of desperate terror I wish to hurtle backward like a scuttling crab into my evolutionary shell, to be swinging ape and leaping Tupaiid; yes, or sleeping reptile on a stream bank—anything but the thing I have become. He who enters history encounters burdens he cannot bear, and traps more subtle than his subtlest thought. No one quite knows how man became concerned with this torrent called history. He contrives it out of his own substance and then calls it an "act of God" when the tanks grind forward over his own body.

A strange animal, indeed: so very quiet when one turns over the mineral-hardened skull in a gravel bed, or peers into that little dark space which has housed so much cruelty and delight. One feels that something should be there still, some indefinable essence, some jinni to be evoked out of this little space which may contain at the same time the words of Jesus and the blasphemous megatons of modern physics. They are all in there together, inextricably intermixed, and this is how the mixture began.

II

There are innumerable detailed questions of dating and of detailed anatomical analysis and interpretation of those scant human remains which through gaps of hundreds of thousands, even of millions of years, enable us to secure brief glimpses of our nameless forerunners. For more than a hundred years—

ever since the theory of evolution became biologically demonstrable—these facts have been accumulating. To catalogue them, to debate their several arguments, would require volumes. It is my intention here merely to select for discussion a few key items which continue to intrigue the educated layman, and which not only may help him to comprehend a few of the mileposts of his long journey but may give him as well a better comprehension of his own nature and the built-in dangers it contains. The moment is topical, for it is within the past few years that discoveries have been made which may drastically change our ideas about our earliest past.

We will begin with a warning: more than 90 per cent of the world's animal life of past periods is dead. Though it flourished in some instances longer than the whole period of human development, somewhere along its evolutionary path one of two things happened. It vanished without descendants or was transformed, through still mysterious biological processes, into something else; just as man is now something quite different from what he was ten million years ago. This leads to the inescapable conclusion that, contrary to popular impression, evolution is not something "behind" us—the impression we may get while staring into museum cases.

So long as life exists on the planet, it is still changing, adjusting, and vanishing as natural forces—and among them we must now count man—may dictate. Thus life is never really perfectly adjusted. It is malleable and imperfect because it is always slipping from one world into another. The perfectly adjusted perish with the environments that created them. It is not really surprising, when one thinks about it, that man, who evolved with comparative rapidity, should be among earth's most dangerous juvenile delinquents.

He is literally compounded of contradictions, mentally and physically. He is at one and the same time archaic and advanced. His body and his mind are as stuffed with evolutionary refuse as a New England attic. Once he comes to accept and recognize this fact, his chances for survival may improve. He has come halfway on a trembling bridge toward heaven, but the human brain in its loss of life-preserving instincts passes also along the brink of sanity. Here is a great poet, John Donne, speaking more than three centuries ago of the power of the human intellect:

> Inlarge this Meditation upon this great world, Man, so farr, as to consider the immensitie of the creatures this world produces; our creatures are our thoughts, creatures that are borne Gyants; that reach from East to West, from Earth to Heaven, that doe not onely bestride all the Sea, and Land, but span the Sunn and Firmament at once; My thoughts reach all, comprehend all. Inexplicable mistery; I their Creator am in a close prison, in a sicke bed, any where, and any one of my Creatures, my thoughts, is with the Sunne, and beyond the Sunne, overtakes the Sunne, and overgoes the Sunne in one pace, one steppe, everywhere.

Man, in short, has, like no other beast, tumbled into the crevasse of his own being, fallen into the deep well of his own mind. Like modern divers in the sacrificial wells of the Maya, he has drawn from his own depths such vast edifices as the Pyramids, or inscribed on cave walls the animals of his primitive environment, fixed by a magic that inhabited his mind. He retreats within and he appears outward. Even the fallen temples of his dead endeavors affect, like strange symbols, the minds of later-comers. There is something immaterial that haunts the air, something other than the life force in squirrel and chipmunk. Here, even in ruin, something drawn from the depths of our being may speak a message across the waste of centuries.

A little while ago I handled a flint knife, from Stone Age Egypt, running my hand over its beautifully rippled surface. A human mind, an artist's mind, whispered to me from the stone. I held the knife a long time, just as in another way I might hold in my mind the sunlit Parthenon, feeling some emanation, some re-entering power deriving from minds long past but flooding my own thought with renewed powers and novelties. This is a part, a mystical part if you will, of man's emergence into time and history.

When he entered into himself, as no other animal on the globe is capable of doing, he also entered the strangest environmental corridor on the planet, one almost infinite in its possibilities, its terrors, and its hopes. It was the world of history, of symbolic thought, of culture. From the moment when the human brain, even in its dim red morning, crossed that threshold, it would never again be satisfied with the things of earth. It would heft a stone and make of it a tool grown from the mind; fire would become its instrument; sails on the invisible air would waft it far; eventually a little needle in a box would guide men to new continents and polar snows. In each case there would also be the aura of magic. The powers would not be what we of today call natural; around them would hover a penumbral mystery drawn from the abysses of the mind itself. Time and the foreknowledge of death would rise also in that spectral light. Of the fears that beset our dawning consciousness, the brown bone on the shores of a vanished lake bed will tell us nothing. It will tell us only how we changed.

III

From whence did we come? Over and over again the scholar is asked this question by those who forget the wounds and changes in the bone. Do they ask upon which continent we first stood dubiously erect? Do they ask from what limb in an ancient forest we first hung and by some idle quirk dropped down into the long grass that first received us? Do they want to know at what point we first asked a question of some wandering constellation in the night sky above our heads? Or from what marsh we first dragged our wet amphibian bodies up the shore? Or from what reptilian egg we sprang? Or from what cell in some far, steaming sea?

No, the question has to be contained and caught within the primate order to which we and all manner of ring-tailed, wide-eyed lemurs, blue-chinned monkeys and enormous apes belong. With these we share certain facets of a common bodily structure that speaks of ancient relationships. In a strange, figurative way there was a time far back along the evolutionary road when all this weird array might be seen to shrink to a single tree shrew, a single ratty insectivore upon a branch. Man, at that moment, was one of many potentials. He was and was not, and likewise all his hairy and fantastic kin. They all quivered there upon that single branch in one frail body—Socrates. Confucius and Gargantua, along with the organ-grinder's monkey.

The student asks you, as a child his mother, "Where did I come from?"

"Son," you say floundering, "below the Cambrian there was a worm." Or you say, "There was an odd fish in a swamp and you have his lungs." Or you say, "Once there was a reptile whose jaw bones are in your ear." Or you try again. "There was an ape and his teeth are in your mouth. Your jaw has shrunk and your skull has risen. You are fish and reptile and a warm-blooded, affectionate thing that dies if it has nothing to cling to when it is young. You are all of these things. You are also a rag doll made of patches out of many ages and skins. You began nowhere in particular. You are really an illusion, one of innumerable shadows in the dying fires of a mysterious universe. Yesterday you were a low-browed skull in the river gravel; tomorrow you may be a fleck of carbon amid the shattered glass of Moscow or New York. Ninety per cent of the world's life has already gone. Perhaps brains will accomplish the work of extinction more rapidly. The pace is stepping up."

"Life," a cynical philosopher once shrewdly observed, "is a supremely illogical business. One can become dark from excess of light." This statement is so directly applicable to the study of human evolution that it ought to preface any survey of our past. At first glance everything is simple. We have a bone here, a skull there. Teeth grow smaller, brains grow larger. The upright posture undoubtedly preceded by perhaps a million years or more the appearance of a face and brain faintly comparable to our own.

Even after the brain began to grow, it was long shielded by a shell of bone as thick as a warrior's helmet. It was as if nature itself was dubious of the survival of this strange instrument, yet had taken steps to protect it. I like to think that with the invention of a brain capable of symbolic thought—and, as an unsought corollary, philosophy—something behind nature rejoiced to look out upon itself. That massively walled brain, even in its early beginnings, had taken life about three billion years to produce. But the future of no invention can be guaranteed. As in the case of other forms of life in the past, extinction may come about some millennia hence from "natural" causes. Or—as we are constantly reminded by our experts—life's most dazzling invention may, through the employment of its own wizardry, soon be able to erase itself completely from the earth, in a matter not of millennia, but of minutes.

For the human brain, magnificent though it be, is as yet imperfect and bears within itself an old and lower brain—a fossil remnant, one might say—geared to the existence of a creature struggling to become human, and dragged with him, unfortunately, out of the Ice Age. This ancient brain, capable of violent and dominant re-emergence under various conditions of stress, contains, figuratively speaking, claws—claws which by now can be fantastically extended.

Life *is* illogical, and if one looks long and steadily at evolution and at man in particular, the illumination provided by fossil skulls can produce, paradoxically, some profound shadows. In the early history of anthropological studies, when we possessed few human remains but much acquaintance with our living relatives in the trees, the story had seemed simpler: somewhere, not too far back in time, an ape had gotten down from his tree, driven to the ground, possibly, by a growing desiccation of the landscape. In time natural selection had altered an arboreal foot for bipedal progression, and the hands, once used by our ancestors for the manipulation of the branches among which they dwelt, were now employed in the exploration and eventual remaking of the world into which they had intruded. There is, indeed, a certain semblance of truth in this account, but with every discovery of the past few decades, the story has had to be modified if not rewritten. Even today, while no reasonable man doubts the reality of human evolution, its precise pathways are hazy, and far gaps in time and space make the exact succession of forms difficult if not impossible to determine.

It is easy, when bones are few, to stick to a single line of ascent or to give a simple version of events. But bones also have their limitations. We cannot trace the living races far into the past. We know little or nothing about why man lost his fur.

Consider the mistakes to which our descendants, a million years forward in time, would be liable in trying to reconstruct, without a single written document, the events of today. To tell the truth, though theories abound, we know little about why man became man at all.

We know as much—and as little—about our own ancestors as we do about some other missing creatures from the geological record. Why, for example, do bats hurl themselves so suddenly upon us, fully formed, out of the Paleocene era? They emerge with comparative rapidity in the dawn of mammalian history. In fact they bear a distant relationship to ourselves. How they became bats and not men is one of those evolutionary problems which involve the interplay of vast and ill-understood forces operating over enormous lengths of time.

The light being thrown upon our history is truly only the dim light of morning. I often think, on taking off my shoes at night, that they clothe an awkward and still imperfect evolutionary instrument. Our feet are easily sprained or injured, and somehow comical. If they had not been reshaped in some long venture on the early grasslands, we would not find it so satisfactory now to brace them artificially in shoes. The sight of them is chastening to pride. The little toe is attractive to the student of rudimentary and vanishing organs: the over-all perspective is a rude palimpsest, a scratched-out and rewritten autobiography whose first anatomical pages were contrived in some arboreal attic.

Among these living shards and remnants of the past, however, it is easy to linger and be lost. I remarked that we know little about why it was necessary to become man at all. There are many parallelisms in the other parts of nature. Complex social life has arisen several times in diverse insect orders. There are flying marsupials as well as flying placental mammals. But man, the thinker, has occurred but once in the three or more billion years that may be the length of life's endurance on this planet.

He is an inconceivably rare and strange beast who lives both within himself and in his outside environment. With his coming came history, the art of the mind imposing itself upon nature. There has been no previous evolutionary novelty comparable to this save the act of creation itself. Man, imperfect transitory man, carries within him some uncanny spark from the first lightning that split the void. He alone can dilate evil by drawing upon the innocent powers contained in nature: he alone can walk straight-footed to his own death and hold the world well lost for the sake of such intangible things as truth and love.

Yet let me suggest once more that we look long and clearly at ourselves, our strange and naked bodies, our evolutionary wounds, tracked as we have been through trees and lion-haunted grasslands and by the growing failure of instinct to guide us well. Let us take care, for beyond this point in time, brains and sympathy—the mark of our humanity—will alone have to guide us. The precedent of the forest will be wrong, the precedent of our dark and violent mid-brains will be wrong; everything, in short, will be wrong but compassion—and we are still the two-fold beast. Why did we have to be man, we ask ourselves, as the Christians of another day must have asked: "How can man be made whole? How can he be restored to the innocence he knew before the Fall?"

The stylist and philosopher George Santayana gave us one of his great insights when he wrote sadly, "The Universe is the true Adam, the Creation the true Fall." He saw that to come out of the blessed dark of nonbeing, to endure time and the disturbances of matter, is to be always subject to the unexpected even if it masquerades as "natural law." With the unexpected comes evil, the unforeseen, the moment's weakness. Life—even nonhuman life—becomes parasitic, devours its fellows, until a Darwin looking on may call it "devil's work." The creation falls and falls again. In mortal time, in Santayana's sense, it must ever fall. Yet the falling brings not only strange, dark and unexpected ends to innocent creatures but also death to tyrannous monsters.

The very novelties of life offer renewed hope to the spirit that works upon intransigent matter and lends us our willingness to endure our time. For us, for this little day we inhabit so unthinkingly, much has been suffered. A gray and shadowy and bestial thing had to become a man. Gleams, strange lights, half-caught visions of both love and abominable terror, must have dogged our footsteps. Disease destroyed us in infancy. We were abruptly orphaned, and great teeth struck us down. We were fearful of the dead who haunted our dreams. We barked and gabbled until, at some unknown point in time, the first meaningful invented sounds in all the world were heard in some lost meadow. The creature had stumbled, with the growth of speech, into a vast interior world. Soon it would dominate his outer world.

A short time ago most of us who work in this graveyard of the past would have said that a brain which we could truly denominate as human was perhaps no older than the lower Ice Age, and that beyond the million years or so of Ice Age time, man, even low-browed, thick-skulled man, had vanished from our ken. If, that far back, he still walked, he was not a tool-user; if he still talked, his thoughts had found no lasting expression upon the objects of his outer world. It appeared to us, not that he had vanished in the seven-million-year epoch of the Pliocene, but rather that he was a thinly distributed ground ape, a late descendant upon the upland grasses, still teetering upon a dubiously adapted foot from one sparse clump of trees to another.

In July 1961 our ideas were destined to change drastically. They were to change not so much because of a newly described form of early man from Africa—we had grown used to that—but rather because of what a new method of dating was to tell us about humanity in general.

IV

Over the previous thirty years a startling series of discoveries in South and East Africa had revealed that the simplified versions of single-line human evolution were very unlikely to be true. It was soon realized that African humanity has a very ancient history—more ancient than, at present, we can demonstrate for any other part of Asia or Europe. I am not now speaking of problematic early relatives of ours such as *Oreopithecus* from the Tuscan coal mines of Italy, but of tool-using creatures walking upon the ground.

Deep in the Olduvai Gorge in Kenya lay hand axes of enormous antiquity. Even more primitive pebble tools were found in various regions in South Africa. Man—some kind of heavy-browed man—had long roved the uplands of that game-filled continent. Darwin's guess that Africa might prove to be the original home of man was taking on renewed interest, even though very ancient remains such as Peking Man had been located in the caves of Choukoutien, and a series of early forms had also turned up in Java. It must be remembered also that the inhospitable desert break between Africa and southwestern Asia has not always existed. In early ages it is likely that freedom of movement between these two regions was far more simple for primitive man and beast than has been true in historic times. Hence, since so much of Asia remains archeologically unknown, it would be premature to decide that Africa alone contains the full story of the human past. That it has provided us with more clues to early human development than any other region, however, it would now be idle to deny.

All through the past few decades the labors of such pioneer scientists as Robert Broom, Raymond Dart, L. S. B. Leakey and J. T. Robinson have succeeded in turning up amid the breccia of ancient cave deposits a hitherto totally unsuspected and apparently cultureless group of ape-men, or perhaps one should say man-apes. Instead of gorilloid, long-fanged creatures lately descended from the trees, such as the early Darwinists would have envisaged, these creatures, of whom numerous remains and several species have been recovered, brought dramatically home to us a largely unsuspected aspect of the human story, anticipated on theoretical grounds only by Darwin's great contemporary Alfred Russel Wallace.

The idea of the gorilloid nature of early man as advanced by many nineteenth-century scientists was not borne out by the newfound fossils. Instead, the bones proved to be those of rather slightly built, erect-walking "apes" with massive molar teeth unaccompanied by projecting canines. In short, the animals turned out to be a rather variable lot of short-faced, small-brained creatures already adapted for walking on their hind feet. Long arguments developed as to whether these creatures of some five hundred cubic centimeters of cranial capacity—roughly akin to the brain size of a modern chimpanzee or gorilla

—could have made crude tools, or at least utilized the long bones of slain animals as clubs or stabbing weapons. This was possible —but difficult to prove.

One thing, at least, had become evident. The man-apes represented not recently arboreal apes but, instead, an unsuspected variety of erect-walking anthropoids the adaptation of whose feet to a ground-dwelling existence was already largely perfected. In Tertiary times large primates had not been confined to the trees. Instead, they had successfully brought their old-fashioned arboreal bodies down onto the grass and survived there—a feat of no mean magnitude. By some quirk of neurological evolution they had acquired an upright posture which had freed the forelimbs from the demands of locomotion. Man bears in his body clear signs of an early apprenticeship in the trees. We now began to suspect, however, that man had served his arboreal apprenticeship much farther back in time than many scholars had anticipated. It also became evident that the number of forms and datings of what soon came to be called the Australopithecine man-apes could only suggest that not all of them were direct human ancestors. These African creatures hinted rather of a variety of early man-apes, not all of whom had necessarily taken the final step of becoming human.

A group of apes had entered upon a new way of life in open park land and grassland. Arboreal apes are not carnivorous; they are primarily vegetarians. But these man-apes, or perhaps I should say *some* of these man-apes, had become killers of game. Their massive jaws, however, are not evidence of this fact. Massive molar teeth may mean only the consumption of certain types of uncooked vegetation. It is the broken bones of animals in the caves they frequented which suggest that some species, at least, had become killers, using their unloosed forelimbs as weapon-wielders. As for the brain, though it was perhaps still small, the upright posture had given this organ some qualitative advantage over the brains of our living relatives, the great apes.

Still, we had to look upon such a creature as essentially an odd, humanlike ape. Like any other animal, it had intruded into and adapted itself to a grassland existence: that it could speak seemed unlikely. It appears unlikely, also, that all such creatures survived to become men; some may have been living fossils in their own time. The last of them may have been exterminated by the spread of man himself. But they indicate that the bipedal apes were well adapted to survive upon the ground without entering extensively upon a second road of conquest.

It remained for the direct human ancestors, from whatever bipedal group they may have sprung, to precipitate the final stage in man's development: the rise of the great brain, still marked by its ferocious past. For man entered, with the development of speech and its ever-growing product, culture, into the strangest and most rapidly changing environment on the planet, an environment limited only by his own creativeness. He entered into himself; he created society and its institutions. The exterior, natural world would be modified and pushed farther and farther back by the magic circle in which he had immured himself. Some societies would dream on for millennia in a world still close to nature; other roads would lead to the Greek thinkers and the Roman aqueducts. The history of the world-changers had begun.

We can still ask of this varied group of fossils, why did man have to be? No answer comes back. He did not have to be any more than a butterfly or a caterpillar. He merely emerged from that infinite void for which we have no name.

In 1959 L. S. B. Leakey found at Olduvai the massive-jawed, small-brained creature who has come to be known as *Zinjanthropus.* The creature would appear to be not too distant in its anatomy from some of the known, and possibly much later, man-apes. It, however, is remarkable for two reasons. First, it was found in association with clearly shaped stone tools, long known but never found in direct contact with human remains. Thus this creature was not merely a user of chance things which he picked up; he was a thinker who shaped. Second, in 1961 J. F. Evenden and Garniss Curtiss of the University of California announced that *Zinjanthropus* was nearly *two* million years old. They had dated the creature by a new "clock" involving the use of potassium-argon radioactive decay. If this dating method is correct, the history of tool-using man will thus have been carried back almost a million years before the Ice Age—and later Dr. Leakey reported another find, *Homo habilis,* from the same vicinity. The latter he believed to be closer to the human line of ascent than *Zinjanthropus.*

I have mentioned that man's mental development, so far as its later, bigger-brained phase is concerned, has seemed rapid. Dr. Leakey's finds can be interpreted in two ways: as suggesting that the incipient steps leading to the emergence of the large brain began earlier than we have anticipated, or that man drifted in a static fashion on this simple level for a long period before some new mutation or latent dynamism generated a new leap forward in brain size. Little in the way of advanced cultural remains is known before the later Pleistocene, so that the appearance of these tool-using creatures of such archaic countenance is an amazingly disturbing element to our thinking.

Have all our lower Ice Age discoveries been underestimated as to time? And what of the other, the seemingly later yet more primitive Australopithecines? Are they, then, true cousins rather than ancestors—survivals of an even more remote past? We do not know. We know only that darkest Africa is not dark by reason of its present history alone. Contained in that vast continent may be the secret of our origin and the secret of the rise of that dread organ which has unlocked the wild powers of the universe and yet taught us all we know of compassion and of love.

Those ancient bestial stirrings which still claw at sanity are part, also, of that dark continent we long chose to forget. But we do not forget, because man in contemplation reveals something that is characteristic of no other form of life known to us: he suffers because of what he is, and wishes to become something else. The moment we cease to hunger to be otherwise, our soul is dead. Long ago we began that hunger: long ago we painted on the walls of caverns and buried the revered dead. More and more, because our brain lays hold upon and seeks to shape the future, we are conscious of what we are, and what we might be. "No man," wrote John Donne, "doth exalt Nature to the height it would beare." He saw the great discrepancy between the dream and the reality.

Great minds have always seen it. That is why man has

survived his journey this long. When we fail to wish any longer to be otherwise than what we are, we will have ceased to evolve. Evolution has to be lived forward. I say this as one who has stood above the bones of much that has vanished, and at midnight has examined his own face.

Notes
Acknowledgments
Index

Notes

Notes to parts I and IV are by Loren Eiseley; those to Part II by Edward Blyth and those to Part III by Arthur Grote, except where marked L.E.

The papers by Blyth published in *The Magazine of Natural History* in 1835, 1836, and 1837 are referred to after their first listing only by author's name and date; page numbers are those of the original publication.

Sources frequently appearing throughout the notes are indicated after their first listing by the following abbreviations.

Autobiography: Nora Barlow, ed., *The Autobiography of Charles Darwin* (London, 1958).

Foundations: Francis Darwin, ed., *Foundations of the Origin of Species* (Cambridge, England, 1909).

JBAS: Journal of the Asiatic Society of Bengal.

LLD: Francis Darwin, ed., *Life and Letters of Charles Darwin* (London, 1888, 3 v.).

MLD: Francis Darwin and A. C. Seward, eds., *More Letters of Charles Darwin* (London, 1903, 2 v.).

MNH: The Magazine of Natural History, London.

N: Nora Barlow, ed., *Charles Darwin and the Voyage of the* Beagle (New York, 1946; contains the rough notebooks kept by Darwin during the voyage).

Origin: Charles Darwin, *The Origin of Species,* 2nd ed. (London, World's Classics ed., 1935).

Variation: Charles Darwin, *Variation of Animals and Plants Under Domestication* (New York, 1868, 2 v.).

PART I

CHARLES DARWIN, EDWARD BLYTH, AND THE THEORY OF NATURAL SELECTION

Pages 45 through 80

1. Nora Barlow, ed., *The Autobiography of Charles Darwin* (London, 1958), p. 120.
2. Nora Barlow, ed., *Charles Darwin and the Voyage of the* Beagle (New York, 1946), p. 263.
3. *Autobiography,* p. 122.
4. H. D. Geldart, "Notes on the Life and Writings of Edward Blyth," *Transactions, Norfolk and Norwich Naturalists Society* 3:38–46, 1879; H. M. Vickers, "An Apparently Hitherto Unnoticed Anticipation of the Theory of Natural Selection," *Nature* 85: 510–511, 1911.
5. Edward Blyth, "An Attempt to Classify the 'Varieties' of Animals . . . ," *The Magazine of Natural History* (London) 8:40–53, 1835. (See pp. 97–111 in this volume.)
6. Edward Blyth, "On the Psychological Distinctions Between Man and All Other Animals . . . , " *MNH* n.s. 1:1–9, 77–85, 131–141, 1837. (See pp. 141–165 in this volume.)
7. For an account of the history of the concept one should consult Conway Zirkle's "Natural Selection Before the 'Origin of Species,' " *Proceedings of the American Philosophical Society* 84:-71–123, 1941. Dr. Zirkle observes very perspicaciously that "the widespread acceptance of teleology made natural selection an unheeded hypothesis" throughout its earlier history.
8. See *The Descent of Man* (Random House ed., n.d.), pp. 785 n. 31; 788 n. 1; 803 n. 30; 807 n. 47. All these notes by Darwin refer to *MNH* n.s. 1, 1837.

   In *Variation of Animals and Plants Under Domestication* 1st authorized American ed., (New York, 1868, 2 v.), Vol. 1, p. 55 n. 75 contains a reference to *MNH* 6, 1833; p. 82 n. 42 refers to *MNH* 4, 1840; p. 115 n. 67 gives a reference to the same journal, 1, 1829; pp. 335–336 n. 8 indicates a reference to *MNH* 8, 1835, which is the precise volume of Blyth's first paper on natural selection. Turning to Vol. 2, we find the following citations: p. 30 n. 47, *MNH* 1, 1837; p. 99 n. 33, *MNH* 1, 1837; p. 188 n. 30, *MNH* 6, 1833; p. 189 n. 39, *MNH* 9, 1836; p. 192 n. 52, *MNH* n.s. 2, 1838; p. 193 n. 57, *MNH* 5, 1832; p. 237 n. 7, *MNH* 8, 1835; p. 354 n. 4, *MNH* n.s. 1, 1837; p. 396 n. 21, *MNH* 1, 1829.

9. In *Variation* Darwin remarks, "Mr. Blyth has freely communicated to me his stores of knowledge on this and all other related subjects" (Vol. 1, p. 164 n. 1). The reference is to Oriental pigeons.
10. *Origin of Species* (London, World's Classics ed., 1935), p. 17.
11. *Variation,* Vol. 2, pp. 30–31.
12. *Life and Letters of Charles Darwin* (London, 1888), Vol. 1, p. 83. Francis Darwin (p. 153) spoke of his father's literary habits as follows: "When collecting facts on a large scale, in earlier years, he used to read through, and make abstracts, in this way, of whole series of periodicals." Writing to Huxley in 1859, Darwin commented, "I have picked up most by reading really numberless special treatises and *all* agricultural and horticultural journals; but it is a work of long years" (*Ibid.,* Vol. 1, p. 281).
13. *N,* p. 244.
14. *Ibid.,* p. 260.
15. *Ibid.,* p. 257.
16. See, for example, James H. Fennel, "Comments on Mr. Blyth's Remarks on Specific Distinctions," *MNH* 9:647–648, 1836. Fennel says, "I do not think that the order of nature has offered any opposition to the structure and habits of any animal becoming . . . gradually altered in a greater or less degree from its primitive parents." Fennel, as is characteristic of the British scientists of this period, finds it necessary to dissociate himself from Lamarck, but his meaning is plain.
17. *N,* p. 263.
18. Its use occurs in Edward Blyth, "Observations on the Various Seasonal and Other External Changes Which Regularly Take Place in Birds . . . ," *MNH* 9:399, 1836, (see pp. 112–140 in this volume), and in Blyth, 1837, pp. 399,508.
19. *Origin,* Ch. 13, p. 386.
20. I have discussed this matter at more length in my book *Darwin's Century* (New York, 1958); see particularly p. 329.

21. William Swainson, *A Preliminary Discourse on the Study of Natural History* (London, 1834), pp. 91–92; Peter Rylands, "On the Quinary, or Natural System of M'Leay, Swainson, Vigors, etc.," *MNH* 9:130–138, 175–182, 1836. See also *More Letters of Charles Darwin*, ed. by Francis Darwin and A. C. Seward (London, 1903, 2 v.) Vol. 1, p. 305 n. 1.
22. Blyth, 1836, p. 406.
23. Blyth, 1835, pp. 45–46.
24. Blyth, 1837, p. 135.
25. The actual phrase is used. Blyth may have drawn it from Lyell, with whose *Principles of Geology* he was acquainted. See Blyth, 1835, p. 48. The phrase occurs in Blyth, 1835, p. 46.
26. Blyth, 1835, pp. 45–46. The italics are Blyth's.
27. Similar expressions occur in Blyth, 1837, pp. 79–80, 135.
28. *Darwin's Century*, pp. 53–54, 122, 137, 201–202, 329.
29. Blyth, 1837, p. 83.
30. Blyth, 1836, pp. 406–407.
31. Blyth, 1835, p. 53.
32. Blyth, 1837, p. 134. Compare Blyth's sentence with the following remark from *Origin*, pp. 64–65: "When a species, owing to highly favorable circumstances increases inordinately in numbers in a small tract, epidemics—at least this seems generally to occur with our game animals—often ensue. . . ."
33. Blyth, 1837, p. 135. Italics L.E.
34. *Ibid.*, p. 136.
35. *Ibid.*, p. 137. Note, incidentally, that Darwin devoted considerable attention to the effects of sea water on seeds. See *Edinburgh Philosophical Journal* n.s. 4:375–376, 1856.
36. Blyth, 1835, p. 47.
37. See *Foundations of the Origin of Species,* edited by Francis Darwin (Cambridge, England, 1909), p. 247; *Origin,* pp. 174–175.
38. *Origin,* pp. 7–8, 10; *Foundations,* pp. 1–2, 14–78.
39. Blyth, 1835, p. 49.
40. Blyth, 1835, p. 47. It will be remembered that Darwin, in the notebook of 1836, took a sudden interest in tailless cats.
41. *Foundations,* pp. 59–60.
42. *Ibid.*, pp. 216–17 for swine; p. 62 for five-toed fowls.

43. Blyth, 1835, p. 43.
44. *Foundations,* p. 64.
45. Blyth, 1835, p. 48.
46. *Foundations,* pp. 107–108.
47. Blyth, 1835, p. 44.
48. *Foundations,* p. 91; *Origin,* p. 10.
49. Blyth, 1835, pp. 51–53; 1837, p. 80.
50. *Origin,* p. 77.
51. *Ibid.*
52. *Ibid.*
53. One could add smaller items such as Blyth's reference to the variation manifested in the beaks of finchlike birds (1836, p. 400) which may well have intrigued Darwin through the problem presented by the Galápagos finches which he had collected.
54. Blyth, 1837, p. 138.
55. *Foundations,* pp. 19, 119 n.2, 124–125.
56. Blyth, 1837, p. 5.
57. *Foundations,* p. 19.
58. For example, Darwin follows Blyth quite explicitly in his essay of 1842 when he says that if in any country or district all animals of one species are allowed freely to cross "any small tendency in them to vary will be constantly counteracted." "Such varieties," Darwin says, "will be constantly demolished" (*Foundations,* pp. 2–8). Compare Blyth, 1835, pp. 46–47.
59. Charles Lyell, *Principles of Geology,* 3rd ed. (London, 1834, 4 v.), Vol. 2, p. 364.
60. *Ibid.,* pp. 345–346.
61. *Ibid.,* p. 375.
62. *Ibid.,* p. 351.
63. Blyth, 1835, p. 46. Darwin's autobiography states that his interest in the cross-fertilization of flowers by insects, begun in 1838, arose from his belief that "crossing played an important part in *keeping specific forms constant.*" It is worth noting that, here he is treating of the conservative aspect of selection just as it was handled by Blyth in the paragraph quoted.
64. Blyth, 1835, p. 46. The italics are Blyth's.
65. Blyth, 1837, p. 137.

66. *Ibid.*, p. 136.
67. S. A. Barnett, ed., *A Century of Darwin* (London, 1958), p. xiii. W. L. Brown, Jr., has treated another aspect of the break between modern species (character displacement) in several papers. For a discussion and bibliography see his "Some Zoological Concepts Applied to Problems in Evolution of the Hominid Lineage," *American Scientist* 46:151–158, 1958.
68. Blyth, 1835, p. 46. See also Blyth, 1837, p. 80.
69. The term appears to be used by Blyth and Darwin in the sense of adjoining, not blending.
70. *N,* p. 263.
71. As early as the first edition of the *Journal of Researches* (1839), we can observe Darwin's thought playing over this problem of the successful intrusion of foreign types of animals. "I will add," he says at the close of Chapter IX, "one other remark. We see that whole series of animals, which have been created with peculiar kinds of organization, are confined to certain areas; and we can hardly suppose these structures are only adaptations to peculiarities of climate or country; for otherwise, animals belonging to a distinct type, and introduced by man, would not succeed so admirably, even to the extermination of the aborigines. On such grounds it does not seem a necessary conclusion, that the extinction of species, more than their creation, should exclusively depend on the nature (altered by physical changes) of their country" (*JR,* p. 212). Darwin is here already attempting in a subterranean fashion to find a way past the fixity of Blyth's totally adapted world.
72. Lyell, Vol. 2, p. 369.
73. W. D. Weissenborn, "On the Influence of Man Modifying the Zoological Features of the Globe . . . ," *MNH* n.s. 2:13–18, 65–70, 122–128, 239–256, 1838.
74. *Foundations,* p. xvi. The remarks quoted by Francis from Charles' notebook take on an added clarity when read in the light of Blyth's work.
75. A second obstacle to a complete acceptance of Darwin's statement lies in an unnoted half-discrepancy of the *Autobiography* itself. Darwin says he read Malthus in October 1838 and at this point glimpsed how the principle of artificial selection could be transformed into natural selection in wild nature. Yet farther on

in Darwin's own reminiscences (p. 127) he tells us that he was studying the cross-fertilization of flowers with an eye to the species problem as early as the *summer* of 1838. As I noted (pp. 64) this same interest in the conservative aspects of intercrossing is expressed by Blyth. It seems a little unlikely that if Darwin first grasped the Malthusian principle in the fall of 1838 he would have been so intensively occupied with the conservative aspects of crossing in the summer of the same year unless his thought on natural selection was already well advanced. Rather it suggests, once more, that Darwin was engaged in seeking a way through this obstacle to divergence observed by Blyth and of which Darwin was already aware through a perusal of Blyth's paper.

76. Leonard Huxley, ed., *Life and Letters of Sir Joseph Dalton Hooker* (London, 1918), Vol. 2, p. 43.
77. See, for example, his letters to Jenyns in 1845 (*LLD,* Vol. 2, pp. 31–32, 33–35.
78. Gerald Henderson, *Alfred Russel Wallace: His Role and Influence in Nineteenth Century Evolutionary Thought* (Doctoral dissertation, University of Pennsylvania, Philadelphia, 1958), p. 54.
79. *Darwin's Century*, p. 182 n. 17.
80. Blyth, 1837, p. 82.
81. Edward Blyth, "On the Doctrine of Spontaneous Organization," *MNH* n.s. 2:508–509, 1838.
82. Interesting in this connection is a letter from Lyell to A. R. Wallace written in 1867. Lyell says: "When I first wrote, thirty-five years ago, I attached great importance to preoccupancy, and fancied that a body of indigenous plants already fitted for every available station would prevent an invader, especially from a quite foreign province, from having a chance of making good his settlement in a new country. But Darwin and Hooker contend that continental species which have been improved by a keen and wide competition are most frequently victorious over an insular or more limited flora and fauna." James Marchant, *Alfred Russel Wallace: Letters and Reminiscences* (New York, 1916), p. 278.
83. *LLD,* Vol. 1, p. 358.
84. *LLD,* Vol. 2, p. 78.
85. *Foundations,* p. 34 n. 1. Compare with Blyth, 1837, p. 80.
86. Later, in a letter to Lyell, Darwin on the eve of the *Origin* gave

forceful expression to this new historical point of view. He says, "As each species is improved, and as the number of forms will have increased, if we look to the whole course of time, the organic condition of life for other forms will become more complex, and there will be a necessity for other forms to become improved, or they will be exterminated; and I can see no limit to this process . . ." *LLD,* Vol. 2, p. 177.

87. *Foundations,* pp. 30, 35. See also *LLD*, Vol. 2, pp. 209, 259.
88. *Foundations,* pp. 15, 91; Blyth, 1835, pp. 44–45.
89. *Foundations,* pp. 90–91.
90. *Ibid.,* p. 33.
91. *Origin,* p. 156.
92. *LLD,* Vol. 2, p. 78.
93. Patrick Matthew, *On Naval Timber and Arboriculture* (Edinburgh, 1831), p. 308.
94. *Variation,* p. 287.
95. Matthew, p. 107. This is the page reference given by Darwin which covers part of his comment but not all. To get the part referring to the variability of the trees one must go back to my reference to page 308 of Matthew.
96. *LLD,* Vol. 2, p. 301.
97. The *Foundations of the Origin of Species* is deceptive because, as Francis Darwin explains in the preface, he added subheads to the first essay in order to prepare the rough draft for publication. I can find no trace of "natural selection" in the text. It occurs only as a subhead added long afterward by Darwin's son and in one scrap added on the back of a page. This may be later than the text. See *Foundations,* pp. xxi–xxii, 44 n. 4.
98. Matthew (p. 387) also uses the phrase "selection by the law of nature."
99. One curious little episode took place as the *Origin* was about to be published. Apparently Murray, Darwin's publisher, had raised some question about the use of the words "natural selection" in the full title. Darwin wrote to Lyell, "Why I like the term is that it is constantly used in all works on breeding, and I am surprised that it is not familiar to Murray; but I have so long studied such works that I have ceased to be a competent judge . . ." (*LLD,* Vol. 2, p. 153). This odd little statement cannot mean what it appears to mean or Darwin would actually be denying his own originality.

The word "selection" was in common use among breeders but, so far as I have been able to ascertain, none before Darwin were using the expression "natural selection." Darwin himself was to define the phrase in the *Origin* with no such accompanying renunciation of originality as that quoted here. Furthermore, he was later to complain frequently of the public's failure to understand it. See, for example, *LLD,* Vol. 2, pp. 317–318; also *MLD,* Vol. 1, pp. 160–161. It would appear that in the haste of composition he allowed the pronoun "it" to stand for "natural selection" when what he intended to designate was "selection."

100. It is interesting to note that Lyell caught Darwin up in the proofs of the first edition of the *Origin* when he was about to ignore Lamarck and St. Hilaire as evolutionary forerunners (*LLD,* Vol. 2, p. 207). Darwin himself omitted Wallace "by inadvertence" from the final summary of the first edition of the *Origin.* Although this was remedied in later editions, Wallace's name was never afterward in any edition indexed for this particular spot. Doubtless these acts were unconscious and not deliberate but they have a certain psychological interest and consistency. See *LLD,* Vol. 2, p. 264. Of Lamarck's book Darwin wrote to Lyell in October 1859, "It appeared to me extremely poor; I got not a single fact or idea from it" (*LLD,* Vol. 2, p. 215). On another occasion in 1859, shortly after the *Origin* was published, Darwin remarked to Hooker, "I have always had a strong feeling no one had better defend his own priority. I cannot say that I am as indifferent to the subject as I ought to be . . ." (*LLD*, Vol. 2, p. 252).
101. *LLD,* Vol. 1, p. 355; *Autobiography,* p. 124.
102. He writes, in a letter to John Morley in March 1871, "I believe your criticism is quite just about my deficient historic spirit, for I am aware of my ignorance in this line" (*MLD,* Vol. 1, p. 326). Other evidences of Darwin's historical indifference, even to the letters of his most distinguished colleagues, are to be found in his son Francis' account of his filing habits: "When his slender stock of files was exhausted, he would burn the letters of several years. . . . This process . . . destroyed nearly all letters received before 1862" (*LLD,* Vol. 1, p. v). In this manner one of the most important letters Hooker ever wrote to Darwin has perished. See J. R. Baker, "A Critique of Materialism," *The Hibbert Journal* 45:32, 1936.

103. Writing to W. Graham in 1881, Darwin said, "I think I could make somewhat of a case against the enormous importance which you attribute to our greatest men; I have been accustomed to think second, third, and fourth rate men of very high importance, at least in the case of Science" (*LLD,* Vol. 1, p. 316).
104. *LLD,* Vol. 2, pp. 315–316.
105. Alexander Maxwell, *Plurality of Worlds* (1817), p. 178.
106. Swainson, p. 88.
107. *MLD,* Vol. 2, p. 147.
108. Darwin's concern over the intellectual climate of the times can be glimpsed in a letter to Gray as late as 1857. "You will perhaps think it paltry of me, when I ask you not to mention my doctrine; the reason is, if anyone, like the author of the *Vestiges* were to hear of them, he might easily work them in, and then I should have to quote from a work thoroughly despised by naturalists, and this would greatly injure any chance of my views being received by those alone whose opinions I value" (*LLD,* Vol. 2, p. 122).
109. T. R. R. Stebbing, Letter regarding Butler's contention that Darwin slighted the older evolutionists, *Nature* 23:336, 1881.
110. "Purpose and Particles in the Study of Heredity," in *Science, Medicine and History,* ed. by E. A. Underwood (London, 1953), Vol. 2, p. 74. Jacques Barzun in his well-known study *Darwin, Marx, Wagner,* 2nd rev. ed. (New York, 1958), expressed similar views (p. 18) and quotes H. F. Osborn (p. 52) as saying that Darwin "owed far more to the past than is generally believed or than he himself was conscious of. . . ."
111. *Autobiography,* p. 219.
112. "You must remember that I am now publishing an abstract, and I give no references" (*MLD,* Vol. 1, p. 118). The letter was written to Wallace in April 1859.
113. Charles Coy, "The Individuality of Charles Darwin," *Popular Science Monthly* 74:345–346, 1909.
114. See Douglas Hubble, "The Life of the Shawl," *Lancet* 265:1351–1354, 1953; also Rankin Good, "The Life of the Shawl," *Lancet* 266:106–107, 1954.
115. It is worth noting that the biologist William Ritter was struck, a number of years ago, with the fact that Darwin had become convinced of the fact of evolution before natural selection "had

occurred to him." In the light of the present essay Ritter's observation of the dichotomy between belief and discovery takes on renewed interest. "Mechanical Ideas in the Last Hundred Years of Biology," *American Naturalist* 72:318–319, 1938.

116. *MLD,* Vol. 1, p. 475.
117. *MLD,* Vol. 1, p. 63.
118. Francis Darwin says, "His letters to my father give evidence of having been carefully studied . . ." (*LLD,* Vol. 2, p. 316). To my knowledge, these letters have never been published. Nor were Darwin's letters to Blyth forthcoming at the time the published collections were made (*MLD,* Vol. 1, p. 62).
119. In a letter written to Lyell in September 1838 we find Darwin speaking enthusiastically of the "delightful number of new views which have been coming in thickly and steadily, on the classification and affinities and instincts of animals—bearing on the question of species. Notebook after notebook has been filled . . ." (*LLD,* Vol. 1, p. 298). It will be recalled that Blyth's papers dealt with these matters and that Darwin had utilized the journals which contained them.
120. Arthur Grote, "A Memoir of the late Edward Blyth," *Journal of the Asiatic Society of Bengal,* Pt. 2, 43: xiv, 1875 (see pp. 169–183 of this volume).

DARWIN, COLERIDGE, AND THE THEORY OF UNCONSCIOUS CREATION

Pages 67 through 93

1. Gavin de Beer, ed., "Some Unpublished Letters of Charles Darwin," *Notes and Records of the Royal Society of London* 14:-1:53, June 1959.
2. Max Schulz, *The Poetic Voices of Coleridge* (Detroit, 1963), p. 5.
3. *Autobiography,* p. 125.
4. See "Charles Darwin, Edward Blyth, and the Theory of Natural Selection," in this volume.
5. De Beer, ed., "Darwin's Notebooks on Transmutation of Species," *Bulletin of the British Museum (Natural History),* Historical Series, Vol. 2, Nos. 2–5, 1960.
6. *Ibid.,* Second Notebook, Part II (Vol. 2, No. 3, p. 106).
7. Blyth, 1837, p. 147.

8. De Beer, introduction to First Notebook, Part I (Vol. 2, No. 2, p. 36).
9. De Beer, *Charles Darwin: Evolution by Natural Selection* (New York, 1964), p. 102.
10. George Wald, "Innovation in Biology," *Scientific American* 199:-3:100, September 1958.
11. Theodosius Dobzhansky, "Blyth, Darwin, and Natural Selection," *American Naturalist* 93:870:204, 205, May–June 1959.
12. When Dr. Dobzhansky proposed his theory in the *American Naturalist* he very graciously asked me if I would care to make a response. At the time, travel and administrative duties prevented me from accepting Dr. Dobzhansky's invitation. The present article exploring this subject was really stimulated by his generosity, and I am very glad to acknowledge this fact.
13. Dobzhansky, p. 205.
14. *Ibid.*, p. 206.
15. There is a failure here, however, to distinguish between creativity in the arts and sciences and its traditional modes of expression. The poet is not called upon to footnote or to give the history of his ideas. The scientist by tradition honors and cites the significance of his precursors.
16. Schulz, p. 7.
17. John Livingston Lowes, *The Road to Xanadu* (Boston, 1927), pp. 59–60.
18. Werner W. Beyer, *The Enchanted Forest* (New York, 1963), p. 113.
19. *Ibid.*, p. 49.
20. *Ibid.*, p. 47.
21. *Ibid.*, p. 113.
22. *Ibid.*, p. 66, citing R. C. Bald, "Coleridge and *The Ancient Mariner*," *Nineteenth-Century Studies* (Ithaca, N.Y., 1940).
23. Beyer., p. 75.
24. Lowes, p. 228.
25. Beyer, p. 186.
26. He was able, however, to refer to everything else about Blyth's work in detail.
27. S. T. Coleridge, *Anima Poetae,* ed. by E. H. Coleridge (London, 1895), pp. 87–88.

28. Lowes, p. 43.
29. J. B. Beer, *Coleridge the Visionary* (New York, 1962), p. 185.
30. S. T. Coleridge, Notebook XVI, 6–13 December 1803.
31. Schulz, p. 104, citing Lamb.
32. A parallel is seen in the case of Coleridge. A suggestion of plagiarism made by De Quincey concerning his friend Coleridge brought several critical replies. De Quincey, says John Metcalf, was accused of "bad taste, not to say treachery." One Coleridge enthusiast even declared that "one might as well . . . accuse the bee of theft for gathering treasures from many flowers." Sara Coleridge, while admitting her father's plagiarism, pleaded that "if he took, he gave." See John Metcalf, *De Quincey: A Portrait* (New York, 1963), p. 115.
33. C. D. Darlington, *Darwin's Place in History* (Oxford, 1959), p. 57.
34. *Autobiography,* p. 124; George Gaylord Simpson, review in *Scientific American* 199:2:119, August 1958.
35. *Autobiography,* p. 153.
36. Leonard Huxley, ed., *Life and Letters of Thomas Henry Huxley* (New York, 1902, 2 v.), Vol. 2, p. 42.
37. Material for *Variation* was drawn from Darwin's original "big book" of the *Origin.*
38. *Variation,* Vol. 1, pp. 335–336 n. 8.
39. *Ibid.,* Vol. 1, p. 164 n. 1.
40. De Beer, introduction to First Notebook, p. 26.
41. De Beer, First Notebook, p. 41.
42. *LLD,* Vol. 1, p. 129.
43. De Beer, introduction to First Notebook, p. 26.
44. See "Charles Darwin, Edward Blyth, . . ." in this volume.
45. De Beer, introduction to Third Notebook, Part III (Vol. 1, No. 4, p. 126).
46. Huxley, *Thomas Henry Huxley,* Vol. 2, p. 42.
47. *LLD,* Vol. 1, p. 82.
48. *Ibid.,* p. 80.
49. *Autobiography,* p. 125.
50. *LLD,* Vol. 2, p. 109.
51. Simpson review, p. 122.

PART II

THE VARIETIES OF ANIMALS
Pages 97 through 111

1. Roman numerals cited by Blyth represent volume numbers of *MNH.* [L.E.]
2. These observations are chiefly deduced from the results of some experiments with mice and rabbits.
3. Of seven young rabbits thus produced, two were albinoes, one black, and the remainder of the usual color.
4. For some curious remarks on this subject, see the excellent article "Ass" in Partington's *Cyclopaedia of Natural History.*
5. I have not heard, however, that wild bullfinches, hawfinches, and other birds liable to be thus affected, are more commonly found black in localities where hemp is much grown. Amongst others, the skylark and woodlark are very susceptible of the influence of this food.
6. [*A Tame Duck which flies with the same Power, and at the same Height, as a Crow.* (H. S., in I, 378.)—Was not this duck a wild one? I am led to ask this question from having myself witnessed a similar instance. I had often seen a duck, which I had taken to be a tame one, flying about, and always returning to the farm to which it belonged. On inquiry, I found that this duck had been taken, when a duckling, from the nest of a wild duck, and began to fly as soon as it was full grown. The case which H. S. mentions might probably be accounted for in the same manner, as it is by no means likely that so unwieldy a bird as the tame duck should think of trying its wings, after its ancestors had for so many successive generations been satisfied with walking and swimming, and fly "with the same power, and at the same height, as a crow."—W. H. H., Postmark, Burton on Trent, Oct. 8, 1834.

   The late Rev. Lansdown Guilding had remarked as follows on the case stated by H. S.: "Domestic birds, from flying little, have their muscles relaxed, or, perhaps, they never acquire their natural strength, for want of exercise. I have observed the geese in Worcestershire, in harvest time, to take very long flights; but, though they went on boldly, they never ascended very far into the air."—Lansdown Guilding, St. Vincent, May 1, 1830.]
7. Note that here Blyth has already glimpsed the isolating mecha-

nism which was to become, and remain, an important evolutionary topic. [L.E.]

8. See, however, a good practical article on this subject, entitled "Breeding," in one of the forthcoming numbers of the now publishing edition of Miller's *Dictionary of Gardening and Rural Economy.*
9. *Agkōn,* an elbow, from the crooked form of the forelegs. See Lawrence's *Lectures,* pp. 447, 448.
10. See Dr. Stark "on the influence of color on heat and odors," in Jameson's *Edinburgh Philosophical Journal* for July 1834; also Professor Powell's reply to it, in the number for October 1834.
11. This gentleman should have mentioned, in his account of the white stoats seen in summer, whether the tail was white or black. If the former, they were doubtless albinoes; if the latter, some constitutional debility may have prevented them from resuming their natural hues. I have seen white stoats late in March, but never after this. Both in these and in the white ferret (a domestic albino variation of the polecat) a decided tinge of yellow is always more or less noticeable.
12. See Dr. Stark's paper, before cited, in Jameson's *Edinburgh Philosophical Journal* for July 1834. [See *MNH* 6:79.]
13. See Mudie's *Feathered Tribes of the British Islands,* Vol. 1, p. 50.
14. *Ibid.,* Vol. 1, p. 190.
15. Among day-flying Lepidoptera, the more gaudy colors are usually on the *fore* wings.
16. Animals of bright and gaudy colors are generally very retiring in their habits: even the common robin mostly turns away his breast as you approach.

SEASONAL AND OTHER CHANGES IN BIRDS
Pages 112 through 140

1. I am unwilling to allow even this harmless line to pass muster without indulging in a few remarks on the distinctness of the human race from all other parts of the animal creation; a distinctness too little borne in mind by many naturalists. Man alone, of all the countless wonders of creation, though clad in a material

frame, the functions of which are necessarily identical with those of other animals, is no part of the mere reciprocal system of nature; as they are. He alone is bound to no particular locality, but inhabits alike the mountain and the plain, and *by contrivance* is enabled to endure the fervid heats of tropical climes, and the withering blasts of a polar winter; traverses in all directions the wide extent of the pathless ocean, interchanges purposely the productions of distant lands, and accommodates the respective soils for their reception. He alone degenerates in climates which supply his every natural want; and placed as nature formed him, in the richest soil, is a being out of his element, unable, by the mere unassisted use of his own organs, to maintain his existence as a species. He alone studies the complicated laws of matter, that he may wield them at his will. He alone possesses a power of indefinite self-improvement, and can so communicate his attainments that each generation shall rise in knowledge above the last. He alone has the sense to sow, that he may reap; and, alone, intentionally, and from observation and reflection, opposes obstacles to the course of events in their natural progression; reduces whole countries to an artificial state; and systematically increases vastly their capability of yielding sustenance for him, and for those creatures he has taken under his protection. Other races disappear before him, whose existence is at all opposed to his interest, and those alone remain (but oh! how altered from their former condition!) which minister to his wants and comforts. All other beings are mere creatures of locality, whose agency tends to perpetuate the surrounding system of which they are members; but wherever man appears, with his faculties at all developed, the aspect of the surface becomes changed; forests yield to his persevering labors; the marshes are drained, and converted into fertile lands: the very climate accordingly changes under his influence, and oftentimes to the extinction of some of the indigenous products of the soil. Does not, then, all this intimate that the human race is no part of the mere mundane system, that its agency tends rather to supersede, and is opposed to, that of the rest of organic nature? that a time must come, should nought intervene of what in physics we can take no cognizance, when the human race, having peopled all lands, shall have increased beyond the means of subsistence? But alas! who can dive into futurity? The same awful Being who first awakened man into existence, in common with the meanest atom, who appointed his destiny upon earth to be so diverse from that of his other creatures, who endowed him alone with a capacity to reflect upon his Maker's goodness and power, may (I make no appeal here to revelation, writing only in the spirit of natural theology) close his nonconforming career, as a species, upon earth, in a manner different from the extinction of other species which yields to the progressive changes of the surface. No naturalist can doubt that this beautiful world existed, and was clad in verdure, and inhabited, for countless ages before man became its denizen; and there are no memorials to indicate that an analogous being ever previously existed. Man alone is a creature by himself; the only being whose agency is at all opposed to the mutual and reciprocal system of adaptations prevalent around him. He did not always exist here, and there is no reason to suppose that he always will. All conduces rather to intimate that he is but a sojourner for a short time. In his vanity, he is apt to imagine that all were made for him! and presumptiously enquires *of what use* could have been the creation without him! Yet how ardently does he labor to exterminate every portion of that creation, which he deems to be in the least injurious to his own interests!

2. See "Charles Darwin, Edward Blyth . . .," pp. 53–54, 65, in this volume. [L.E.]
3. The "mineral kingdom" is a superfluous epithet, too vague to have any meaning beyond a negative one. Chemically speaking, it, indeed, comprises both the others. The proper distinction is, of course, between *organized* and *not organized.*
4. Curiously enough, however, the song linnet's changes of tint do not, to the slightest extent, ever take place in captivity.
5. Inspection of a considerable number of ptarmigans, at different seasons, induces me to dissent from the general opinion, that the time of molting in these birds is confined to no particular period.
6. I wish the reader to excuse, for the present, my not entering into detail on the moltings of birds, as, just now (this being the chief season for molting), I have some opportunities of considerably extending my information on the subject.
7. On examining a series of specimens of *M. viscivora,* it will be seen that many exhibit conspicuous traces of the mottling on the upper parts, particularly on the rump, and that space covered by the tertiary wing feathers; also on the upper tail coverts; the latter being broadly edged with a paler tint, which in the former occu-

pies the center of each feather. Here we have an interesting illustration, in the plumage of birds, of the gradual development of a particular marking as we recede from the type. There is also a regular increase in the size of the bill, which, in the missel thrush, is rather small. I am unaware that the form of *M. varia* and its immediate congeners is further modified, but suspect them rather to be the extreme ramifications in that direction.

8. From subsequent investigations, I am enabled greatly to strengthen the above position. Minute comparison of a considerable number of American specimens with examples of what have been hitherto esteemed the same species in Europe has brought to light distinctions as curious as, in some instances, they were unexpected. Thus, the osprey of North America may be always told, by trivial though constant characters, from that of Europe; and the same obtains with a variety of other species considered identical.
9. It is greatly to be wished that horticulturists would not name their hybrid plants in the same manner as genuine species; the confusion thus already induced in many genera being quite inextricable. Surely they could find some other mode of denoting them.
10. Since writing this, I have ascertained the fact, that the mule progeny of the *Anser cygnoides,* coupled with the domestic goose, breed freely with one another; and have seen an individual of which both the parents were hybrids. We do not, indeed, know the wild stock of the domestic goose; but, certainly, no one would dream of referring it to *A. cygnoides.* As Mr. Jenyns well observes, the common gander, after attaining a certain age, is always white, a character which, it may be remarked, is in accordance with the snow goose *(A. hyperboreus)* of North America, a species obviously distinct. Let it be, however, borne in mind, that, in every known instance, intermixture of species is solely induced by man's agency; even the mules that have been found wild between *Tetrao tetrix* and *Phasianus colchicus:* for instance, White of Selborne, who figures one of these, states, in one of his first letters, that black game was formerly abundant in the neighborhood, but that only one solitary grey-hen had been seen for many years: such an individual might be expected to breed with a cock pheasant.

PSYCHOLOGICAL DISTINCTIONS BETWEEN MAN AND OTHER ANIMALS

Pages 141 through 165

1. Even more: he will contrive so to place himself, if practicable, that the ferret's eyes shall be dazzled by the light.
2. The indirect effects of human agency on this intuitive knowledge of brutes will be considered presently. In no way is the deterioration more evident, than in domesticated animals poisoning themselves by feeding on that which, in a wild state, they would instinctively reject.
3. The reader may probably be disposed to refer this to the structure of the vocal organs. But, admitting to the full extent the reasonableness of this view, it must be borne in mind that the smaller birds have great power of modulation; and it is a certain fact, that, although in most species the song is purely innate, there are many (as the song thrush and nightingale) in which it is, for the most part, acquired; as is proved by the fact of these never warbling their wild notes when reared in confinement, except they have had opportunities of listening to the proper song of their species; which latter, it may be remarked, they imitate much more readily than any other. I do not consider, however, the music of a bird to be so much the language of its species, as those various notes and calls by which different individuals commune together; and these I have never known to vary under any circumstances.
4. Brutes appear to reason from innate knowledge, and this in proportion to the development of the cerebrum; but the extreme promptitude of their expedients (as will be shown), in cases of emergency, often prohibits us from inferring that these can be the result of aught else than intuitive impulse.
5. I have noticed a remarkable instance of this, on placing down a stuffed polecat before a young brood, tended by a bantam hen. A rail or gallinule will also run towards a bank approximating to their own color; and, if no hiding place be discoverable, will insert the head into a crevice, and, remaining motionless, suffer themselves to be taken. Of this I have known many instances.
6. It is no new remark, that rodents are much below the Carnivora in the scale of intelligence; a necessary consequence of their inferiorly developed brain. Yet few animals have more instinctive cunning and resource than the common rat: but this is not intellect, of which it displays scarcely any when brought up tame; a

condition which, as will be shown, is sure to call forth the non-instinctive intelligence of animals. Judging from my own observation, I should say that the rat was mentally superior to the house mouse, but inferior to the squirrel; which, in its turn, must yield in intellect to the hare; and, I believe the comparative structure of their brains will be found in accordance.

7. As in the contests of animals for the other sex; whence it follows that the breed is chiefly transmitted by the most stout and healthy.
8. Propensities are similarly transmitted in the human race, but certainly not the knowledge of how these are to be gratified. It is true, however, that our observation in these matters is too much confined to cultivated, domesticated man, who is, consequently, farthest removed from the brute creation. The Australian savages are known to have a great penchant for snails and caterpillars; and I have somewhere read of one of these who had been brought up in a town, and carefully kept away from all communion with others of his race, who nevertheless exhibited the same fondness for these dainties, despite the abhorrence with which all his companions regarded them. His *goût* for them must thus unquestionably have been hereditary; though it is probable he may have learned the fact of their being eaten by his race, which, likely enough, induced him to taste and try them.
9. The reader will observe that the doctrine here controverted is but an application of the exploded hypothesis of M. Lamarck.
10. This paragraph is a very clear and precise statement of the conservative aspect of natural selection as it was seen by Blyth. [L.E.]
11. Since writing this, I have been informed of a solitary instance of a male goldfinch mule producing offspring with a hen canary.
12. A friend informs me that he has repeatedly noticed, in Aberdeenshire, the pairing of a black crow with an ordinary individual of *C. cornix;* and he further assures me that, to judge from its most commonly sitting, the former was in every instance the female bird. (Are not the black individuals noticed in Ireland, and assumed to be *C. corone,* in reality varieties of *C. cornix?*) It may be added, that the circumstances occasioning the alleged union, stated by Temminck, betwixt the *Motacilla lugubris* and *M. alba* require much additional investigation.
13. There is no occasion, here, to follow out all the causes which

combine to bring about the extirpation of species; but I will mention one which appears not to have been duly considered by those who have written on the subject. We have every reason to believe that the original germ of an animal may be developed into either male or female; and it is certain, that external circumstances exercise a very considerable influence in determining the sex of the future being. Now, the results of experiments instituted on sheep by the Agricultural Society of Séverac fully warrant the conclusion, that, where species exist under circumstances favorable for their increase, a greater number of that sex is produced, which, in polygamous animals is most effectual for their multiplication; whereas the contrary obtains, probably, in proportion to the difficulty of obtaining a livelihood. The relative age and constitutional vigor of the parents is likewise an important element in this problem; and, combined with the former, will enable us to calculate an average with tolerable precision. I have collected some very curious facts bearing upon this subject, some of which are extremely difficult of explanation. Mr. Knapp, in his *Journal of a Naturalist,* has the following, which is worthy of close attention: "The most remarkable instance," he observes, of variation in the relative proportion of the sexes, "that I remember of late, happened in 1825. How far it extended I do not know; but, for many miles round us, we had in that year scarcely any female calves born. Dairies of forty or fifty cows produced not more than five or six; those of inferior numbers in the same proportion; and the price of female calves for rearing was greatly augmented. In a wild state," he justly observes, "an event like this would have considerable influence upon the usual product of some future herd" (note to p. 138). This occurred in Gloucestershire. The character of the preceding season is not stated; but, most probably, it was one of scarcity to the parent animals. The following list exhibits the proportion of the sexes in the annual produce of generally six cows, of the Ayrshire breed (four being the same individuals throughout, the remainder their produce), kept in a park in this neighborhood. It commences with the year in which the present superintendent took charge of the stock; and there is no question but that, if the stock-books of other persons who have the care of cattle were to be duly looked over for a series of years, many similar and equally interesting facts would be brought to light.

In 1826 from 6 cows, were born 6 male calves, 0 females.

1827	″	6	″	″	″	6	″	″	0
1828	″	6	″	″	″	6	″	″	0
1829	″	5	″	″	″	4	″	″	1
1830	″	6	″	″	″	3	″	″	3
1831	″	5	″	″	″	0	″	″	5
1832	″	5	″	″	″	0	″	″	5
1833	″	6	″	″	″	0	″	″	6
1834	″	6	″	″	″	0	″	″	6
1835	″	6	″	″	″	3	″	″	3
1836	″	6	″	″	″	2	″	″	4

Thus it appears that, for the first four years, but one female calf was produced out of twenty-three births; that in the succeeding year the proportions were equal; that in the next four years, out of twenty-two births, there was not a single male; and that in the following year, again, the sexes were in like proportions. The present season, alone, has formed an exception to this remarkable regularity, which I have in vain endeavored to solve by making every inquiry concerning the male parents. There is some reason, also, to suspect that the same phenomenon will be found to obtain among wild birds. The Hon. and Rev. W. Herbert remarks, incidentally, that he has found in the nests of whitethroats *(Curruca cinerea)* a great predominance of males, and the contrary in those of whinchats and stonechats; which latter I have also noticed myself; but cannot say that I have remarked it in a sufficient number of instances, nor over a sufficient extent of ground, nor for a sufficiently protracted period, to be enabled to deduce any general or satisfactory conclusion: the fact can, in most instances, be only ascertained (without slaughtering a great number) by raising them to maturity in confinement. But the young stonechat may be readily distinguished even in the nest: the immature males have a large pure white spot above their wings, which in the females is pale brown. The subject is extremely worthy of further investigation, and it is needless to point out its important bearings in wild nature.

14. It is amusing to observe how gravely the loss of these parasites is commented on in Vol. IX, p. 612, as a necessary consequence of the extermination of human beings. Let us suppose they were to perish; what then? Have not myriads upon myriads of every class of beings become extinct, as species, without affecting at all

the workings of the mighty system? Why, then, should the *dreaded* loss of a few parasites, the sphere of whose influence cannot be supposed to extend beyond that of the species to which their adaptations link them?

15. The direct influence of decline of temperature in prompting the equatorial movements of the feathered race, may be observed in the fluctuations in intensity of the erratic impulse, throughout the greater part of winter, exhibited by migrant birds in a state of confinement; such variations being constantly found to accord with thermometrical changes. It may be added, that the degree of temperature which incites them to migrate varies considerably in different species; and in some instances, also, it must not be concealed, that the impulse to quit the breeding station is entirely independent of decrease of temperature; as is exemplified by the swift and adult cuckoo retiring southward at the hottest period of the year: so powerful, too, is this impetus in the first-named species, and others of the Hirundinidae, that these have been many times known to leave a brood of half-fledged nestlings to perish. As regards the polar movement, the proximate cause will appear on consideration of the following facts: It is known that, in the feathered race, the enlargement of particular organs in spring superinduces, in most groups, some considerable change in the external aspect; frequently altering, for instance, the color of the bill, and occasioning (in single-molting species) the rapid disappearance of those deciduous edgings to the feathers, which oftentimes conceal, for a while, the brighter tints of summer; which latter, also, are, in addition, commonly more or less heightened at this period. Now, all these changes are observable in two nearly allied species, the chaffinch and bramble finch, both of which pass the winter in the same localities; but it uniformly happens that the vernal change takes place in the former species several weeks earlier than in the latter. In the beginning of March, every chaffinch is found to exhibit its complete summer aspect; whereas, late in April, I have watched, with a glass, a flock of bramble finches feeding on elm blossoms, in none of which had the bill acquired its blue color; coincident with which change this species always leaves the country. The fact is equally noticeable when they are kept in confinement. Fieldfares and redwings, also, linger in our fields till long after their resident congeners have been engaged in breeding; and it is found, on dissecting these, at this period, that they are comparatively very backward in their

seasonal developments, the attainment of which immediately prompts the migrative impulse. Of course, the breeding station is the proper home of a species, and thereto do all its adaptations directly refer; and thus we find that even the genial influence of a more equatorial abode fails to excite the breeding energies of migrant birds, until such time as their distant summer haunts become fitted for their reception. To conclude this subject, it may be added, that the migratory restlessness in caged birds does not dissipate in spring, at the time of the reappearance of their wild brethren, but is occasionally evinced throughout the summer, till its cessation follows the decrease of those organs which had all along stimulated its manifestation; a constitutional change which likewise puts a stop to song, and brings about the autumnal renovation of plumage.

16. St. Helier's, Jersey.
17. Such is, at least, the uniform result of my experience; though I could never discern a corresponding difference in the adults. This curious fact was first intimated to me by a person who had a number of young larks for sale, among which were two nests of very rufous birds, and three of a much darker color: the former, he assured me, were found in a gravelly situation; the others on a dark soil. Some cases I have since noticed have verified the observation. On another occasion, I may probably bring together a number of analogous facts, in the form of a paper; but it would occupy too much space to do so here. It may, however, be added, that the agency of many species confers a reciprocity of adaptation; thus, the mode in which sheep graze has a decided tendency to reduce a country to that bare and bleak state which suits best with their healthy condition. Hence would accrue a necessary return of varieties to their normal characters.
18. Individuals of very diverse breeds mostly do so: where the parents more nearly approximate, the young often entirely resemble one or the other.
19. Here the very remarkable fact, however, is not to be overlooked, that the solitary African species of trogon presents a combination of those colors and markings which uniformly distinguish apart its numerous congeners in the Oriental isles from those of South America.
20. This observation is, however, intended to apply merely to those of inland plants; for some maritime species, as the Pandaneae and

cocoa palms, have their seeds encased in sea-proof coverings, especially adapted for floating uninjured on the waves: the restricted distribution of such vegetables is provided for on another principle.

21. Even here it might be objected that man's influence could alone have brought these species into contact; so hard is it to disentangle ourselves entirely from the meshes of human interference. Such an objection would, however, in this instance, be frivolous.
22. "Homo, naturae minister et interpres, tantum facit et intelligit quantum de naturae ordine re vel mente observaverit; nec amplius scit aut potest."—Lord Bacon.
23. Except man shall have domesticated some of these, and artificially transferred them to new localities.

PART III

EDWARD BLYTH
Pages 169 through 183

1. *Proceedings of the Zoological Society*, July 28, This was an "Amended List" of species, of which he had enumerated nine in a summary monograph in the previous February. This paper was reprinted in Taylor's *Magazine of Natural History* in 1841, and again with additional matter in *JBAS* 10:2:858.
2. *JBAS* 15:51.
3. *Proceedings of the Zoological Society,* 1841, 63; 1842, 93.
4. *JBAS* 13:2:51.
5. *Ibid.*, 14:2:cvi.
6. *Ibid.*, 17:1:10.
7. *Ibid.*, 17:2:122.
8. Our common friend Robert Frith, whose name is of frequent occurrence in the curator's reports.
9. *JBAS* 25:237.
10. See note to *My Scrap Book or Rough Notes on Indian Oology and Ornithology,* No. 1, 181.
11. *JBAS* 29:82.
12. *Ibid.*, 31:60.

13. *Ibid.*, 31:430.
14. The Council's action in anticipation of the vote of a meeting was cordially approved at our annual meeting of 1863, but was protested against as illegal by Mr. Oldham.
15. *JBAS* 32:32; 33:73.
16. *JBAS* 33:582. Blyth's catalogue of Mammalia was published in 1863, its last sheets being carried through the press by his friend Jerdon.
17. *Proceedings,* Geological Society of Dublin, Jan. 13, 1864, p. 173.
18. Among the papers left by Blyth is one headed "Origination of the Various Races of Man," which he may have intended to form part of the book here referred to. It contains nothing original, but brings together numerous points of resemblance and contrast observable in the several groups of the order Primates.
19. The date of capture is erroneously given, both by Mr. Blyth and by Dr. Anderson in his cited communication to the Zoological Society.
20. John Gould, *Birds of Asia,* Pt. XXVI, *Trochalopteron blythii.*

PART IV

NEANDERTHAL MAN AND THE DAWN OF HUMAN PALEONTOLOGY

Pages 187 through 200

1. H. W. Janson, *Apes and Ape Lore in the Middle Ages and the Renaissance* (London, 1952).
2. Arthur O. Lovejoy, *The Great Chain of Being* (Cambridge, Mass., 1936).
3. Joseph Ritson, *An Essay on Abstinence from Animal Food* (London, 1802).
4. *Ibid.*, p. 23.
5. Anonymous, "The Wild Man of the Woods," *Chambers' Journal* ser. 36:129–131, 1856.
6. D. Schaaffhausen, "On the Crania of the Most Ancient Races of Man," *Natural History Review.* 1:155–172, 1861.
7. C. C. Blake, "On the Alleged Peculiar Characters and Assumed

Antiquity of the Human Cranium from Neanderthal," *Anthropological Review.* 2:cxli, 1864.

8. C. C. Blake, "On Certain Simious Skulls," *Report of the British Association for the Advancement of Science,* 1865, p. 114.
9. Rudolph Virchow, *Freedom of Science in the Modern State* (London, 1878), pp. 58–61.
10. Huxley, *Thomas Henry Huxley,* Vol. 1, p. 286.
11. Carl Vogt, *Lectures on Man* (London, 1864); Richard Whately, *Miscellaneous Lectures and Reviews* (London, 1861), p. 304.
12. L. C. Eiseley, "The Reception of the First Missing Links," *Proceedings of the American Philosophical Society..* 98:453–465, 1954.
13. H. B. Tristram, "Recent Geographical and Historical Progress in Zoology," *Contemporary Review* 2:124, 1866.
14. Rudolf Schmid, *The Theories of Darwin and Their Relation to Philosophy, Religion and Morality* (Chicago, 1883), pp. 90–91.
15. W. J. McGee, "Anthropology at the Louisiana Purchase Exposition," *Science* n.s. 22:811–826, 1905.

THE INTELLECTUAL ANTECEDENTS OF *THE DESCENT OF MAN*
Pages 201 through 219

1. St. G. Mivart, "Ape Resemblances to Man," *Nature* 3:481, 1871.
2. J. Burroughs, "The Literary Value of Science," *Macmillan's Magazine* 54:184–191, 1886.
3. J. Needham, "Human Law and the Laws of Nature in China and the West," *Hobhouse Memorial Lectures 1941–1950* (Oxford, 1952), pp. 21–22.
4. B. Willey, "How the Scientific Revolution of the Seventeenth Century Affected Other Branches of Thought," in *The History of Science* (Glencoe, Illinois, 1951), p. 96.
5. E. A. Burtt, *The Metaphysical Foundations of Modern Physical Science,* 2nd rev. ed. (New York, 1951), p. 210.
6. Quoted in W. E. Houghton, *The Victorian Frame of Mind* (New Haven, 1957), pp. 74–75.
7. M. Millhauser, *Just Before Darwin: Robert Chambers and the* Vestiges (Middletown, Conn., 1959), p. 125.

8. H. B. Tristram, "Recent Geographical and Historical Progress in Zoology," *Contemporary Review.* 2:120, 1866.
9. *LLD,* Vol. 1, p. 298.
10. *Ibid.,* Vol. 2, p. 37.
11. *MLD,* Vol. 1, p. 120.
12. *Ibid.,* p. 194.
13. *Ibid.,* p. 184.
14. *Ibid.,* p. 126.
15. *Ibid.,* p. 270.
16. R. Mudie, *A Popular Guide to the Observation of Nature* (London, 1832), pp. 366–371.
17. E. S. Brown, *Probability and Scientific Inference* (London, 1957), p. 11.
18. A. Ellegard, "Darwin's Theory and Nineteenth Century Philosophies of Science," in *Roots of Scientific Thought: A Cultural Perspective,* ed. Philip Wiener and Aaron Noland (New York, 1957), p. 565.
19. C. J. Schneer, *The Search for Order* (New York, 1960), p. 359.
20. J. Bronowski, *The Common Sense of Science* (Cambridge, Mass., 1953), p. 88.
21. W. Heisenberg, *The Physicist's Conception of Nature* (London, 1958), p. 42.
22. K. F. Gantz, "The Beginning of Darwinian Ethics 1859–1871," in *The University of Texas Studies in English* (Austin, 1939), p. 185.
23. *Ibid.,* p. 184.
24. J. H. Tufts, "Darwin and Evolutionary Ethics," *Psychological Review* 16:198, 1909.
25. W. S. Quillian, Jr., *The Moral Theory of Evolutionary Naturalism* (New Haven, 1945), p. 136.

Acknowledgments

Grateful acknowledgment is made to the following publishers and institutions which have given permission for the use of articles and illustrations. Some changes have been made in the original text, with the approval of the Estate of Loren C. Eiseley, to prepare the articles for book publication.

ARTICLES

"Charles Darwin" by Loren Eiseley was originally published in *Scientific American*, Vol. 194, No. 2 (February 1956), pp. 62–72. Copyright © 1956 by Scientific American, Inc. All rights reserved.

"Alfred Russel Wallace" by Loren Eiseley was originally published in *Scientific American*, Vol. 200, No.2 (February 1959), pp. 70–84. Copyright © 1959 by Scientific American, Inc. All rights reserved.

"Charles Lyell" by Loren Eiseley was originally published in *Scientific American*, Vol. 201, No.2 (August 1959), pp. 98–106. Copyright © 1959 by Scientific American, Inc. All rights reserved.

"Charles Darwin, Edward Blyth, and the Theory of Natural Selection" by Loren Eiseley was originally published in the *Proceedings of the American Philosophical Society*, Vol. 103, No. 1 (February 1959), pp. 94–114. The manuscript was submitted to the Society September 4, 1958.

"Darwin, Coleridge, and the Theory of Unconscious Creation" by Loren Eiseley was published in *Daedalus*, Vol. 94, No. 3 (Summer 1965), pp. 588–602. Copyright © 1965 *Daedalus*. A version of this article appeared originally in *The Library Chronicle*, Vol. XXXI, No. 1 (Winter 1965).

"The Varieties of Animals" by Edward Blyth was originally published in *The Magazine of Natural History*, Vol. 8 (1835), pp. 40–53, Art. IV, under the title "An Attempt to classify the 'Varieties' of Animals, with Observations on the marked Seasonal and other Changes which naturally take place in various British Species, and which do not constitute Varieties." This and the following articles by Blyth were reprinted as an appendix to "Charles Darwin, Edward Blyth, and the Theory of Natural Selection."

"Seasonal and Other Changes in Birds" by Edward Blyth was originally published in *The Magazine of Natural History* in two parts. The first part appeared in Vol. 9 (1836), pp. 393–409, Art. I, under the title "Observations on the various seasonal and other external Changes which regularly take place in Birds, more particularly in those which occur in Britain; with Remarks on the great Importance in indicating the true Affinities of Species; and upon the Natural System of Arrangement"; the article was concluded in Vol. 9 (1836), pp. 505–514, Art. I, under the title "Further Remarks on the Affinities of the feathered Race; and upon the Nature of Specific Distinctions." Although less directly pertinent to the evolutionary problem, than the papers of 1835 and 1837, this article has been included as bearing in a general way upon Blyth's taxonomical views and powers of observation.

"Psychological Distinctions Between Man and other Animals" by Edward Blyth was originally published in *The Magazine of Natural History*, Vol. 1 (new series) (1837), pp. 1–9, Art. I, under the title "On the Psychological Distinctions between Man and all other Animals; and the consequent Diversity of Human Influence over the inferior Ranks of Creation, from any mutual and reciprocal Influence exercised among the Latter"; it was continued in Vol. 1 (new series) (1837), pp. 77–85, Art. IV, and concluded in Vol.1 (new series) (1837), pp. 131–141, Art. VI.

"Edward Blyth" by Arthur Grote was originally published in the *Journal of the Asiatic Society of Bengal*, Part II, new series, Vol. 43, xiv (August 1875), under the title "A memoir of the late Edward Blyth" and included a list of the titles of Blyth's published writings. The memoir was reprinted without the bibliography as an appendix to "Charles Darwin, Edward Blyth, and the Theory of Natural Selection."

"Neanderthal Man and the Dawn of Human Paleontology" by

Loren Eiseley was originally published in *The Quarterly Review of Biology*, Vol. 32, No. 4 (December 1957), pp. 323–329. Copyright © 1957 *The Quarterly Review of Biology.* It was presented in a symposium commemorating the hundredth anniversary of the discovery of Neanderthal man, held at the 123rd meeting of the American Association for the Advancement of Science in New York City on December 27,1956.

"The Intellectual Antecedents of *The Descent of Man*" by Loren Eiseley was originally published in *Sexual Selection and the Descent of Man, 1871–1971*, edited by Bernard Campbell, (Chicago: Aldine Publishing Co., 1972). Copyright © 1972 Aldine Publishing Co.

"The Time of Man" by Loren Eiseley was originally published in *Horizon*, Vol. 4, No. 4 (March 1962), pp. 4–11. Copyright © 1962 American Heritage Publishing Co., Inc. It was reprinted in *The Light of the Past: A Treasury of Horizon* (New York: American Heritage Publishing Co., Inc., 1965).

ILLUSTRATIONS

The American Museum of Natural History: 1, 2, 3, 5, 6, 8, 10, 11, 13, 14

Memories of Ninety Years by E.M. Ward, published by Hutchinson and Company: 7

Picture Collection, The New York Public Library, Astor, Lenox and Tilden Foundations: 9

Prints Division, The New York Public Library, Astor, Lenox and Tilden Foundations: 4, 12

Index